W9-DDO-527

WITHDRAWN FROM LIBRARY

Second Edition

Developing a Safety and Health Program

MONTGOMERY COLLEGE
ROCKVILLE CAMPUS LIBRARY
ROCKVILLE, MARYLAND

Second Edition

Developing a Safety and Health Program

Daniel E. Della-Giustina

CRC Press is an imprint of the
Taylor & Francis Group, an **informa** business

14500125
SEP 27 2011

CRC Press
Taylor & Francis Group
6000 Broken Sound Parkway NW, Suite 300
Boca Raton, FL 33487-2742

© 2010 by Taylor and Francis Group, LLC
CRC Press is an imprint of Taylor & Francis Group, an Informa business

No claim to original U.S. Government works

Printed in the United States of America on acid-free paper
10 9 8 7 6 5 4 3 2 1

International Standard Book Number: 978-1-4398-1438-3 (Hardback)

This book contains information obtained from authentic and highly regarded sources. Reasonable efforts have been made to publish reliable data and information, but the author and publisher cannot assume responsibility for the validity of all materials or the consequences of their use. The authors and publishers have attempted to trace the copyright holders of all material reproduced in this publication and apologize to copyright holders if permission to publish in this form has not been obtained. If any copyright material has not been acknowledged please write and let us know so we may rectify in any future reprint.

Except as permitted under U.S. Copyright Law, no part of this book may be reprinted, reproduced, transmitted, or utilized in any form by any electronic, mechanical, or other means, now known or hereafter invented, including photocopying, microfilming, and recording, or in any information storage or retrieval system, without written permission from the publishers.

For permission to photocopy or use material electronically from this work, please access www.copyright.com (http://www.copyright.com/) or contact the Copyright Clearance Center, Inc. (CCC), 222 Rosewood Drive, Danvers, MA 01923, 978-750-8400. CCC is a not-for-profit organization that provides licenses and registration for a variety of users. For organizations that have been granted a photocopy license by the CCC, a separate system of payment has been arranged.

Trademark Notice: Product or corporate names may be trademarks or registered trademarks, and are used only for identification and explanation without intent to infringe.

Library of Congress Cataloging-in-Publication Data

Della-Giustina, Daniel.
Developing a safety and health program / Daniel E. Della-Giustina. -- 2nd ed.
p. cm.
Includes bibliographical references and index.
ISBN 978-1-4398-1438-3 (hardcover : alk. paper)
1. Industrial safety--Management. 2. Industrial hygiene--Management. I. Title.

T55.D43 2010
658.3'82--dc22 2009042043

Visit the Taylor & Francis Web site at
http://www.taylorandfrancis.com

and the CRC Press Web site at
http://www.crcpress.com

Dedicated to My Grandchildren:

Daniella, Robyn, Katey, James, Daniel Jr., Steven,
Denise, Marissa Lynn, Elisa, and Dana

Contents

Preface

Some things have changed since the first edition of this book. This second edition has been written in response to the reception of the first edition and expanded with review questions at the ends of the chapters to test the reader's comprehension. This publication will enable you to look at the different elements of a Safety and Health program and incorporate them into your organization's efforts, or to aid in developing a program if one is not already in place.

A basic knowledge of these elements is important. Each year a large part of a company's resources (people, property, and efficacy) are lost by not developing and managing a program to protect from those losses that do occur. By putting these key elements to work, hopefully the organization's resources can be preserved.

An understanding of regulations and regulatory bodies is needed to define the requirements for a Safety and Health program and for organizations to achieve compliance. Chapter 1 helps to define what a company's safety policy is and gives an overview of the Occupational Safety and Health Administration (OSHA) and other regulatory agencies.

Hazard communication has been one of the most frequently cited OSHA violations. Chapter 2 will help in hazard communication training and proper writing of material safety data sheets. The chapter will also go into detail about hazardous materials handling and the Chemical Transportation Emergency Center (CHEMTREC).

Knowing the hazards of a job is very important. Chapter 3 on Job Safety Programs will aid in addressing job (workers') safety. The chapter will explain job safety analysis (JSA) procedures to help make tasks safer for the workers. The chapter will also focus in on work-specific safety training and work observations and will touch on the subject of record keeping.

Planned accident investigation will help in being prepared to determine causative factors that might have led to the incident. Chapter 4 gives some suggestions to help assist with accident investigation.

Safety committees can be very helpful in getting employees involved in a safety and health program. The employees who work the jobs every day have a good idea of the safety issues within a particular task and can aid in correcting the problems. This is discussed in more detail in Chapter 5.

Protection from losses and casualties that a fire can cause is very important to any organization. Fires not only can cause a great financial loss but can also lead to the loss of life. Chapter 6 discusses Fire Loss Control Programs along with other responsibilities to maintain a good fire loss control program.

Having a good site emergency response plan is important to prevent mass confusion and to be ready in case a major incident were to occur. Chapter 7 gives some procedures that can be helpful in developing such a plan.

Lockout/tagout procedures play an important part in protecting workers from energized sources. Workers need to be trained on the different aspects of lockout/

tagout rather than job-specific or awareness-level training. Chapter 8 explains how these procedures could prevent the loss of life or severe injury.

Like lockout/tagout, confined space entry is another element where workers need careful and extensive training. Chapter 9 will help in developing a confined space entry program.

Personal protective equipment is used by many companies to protect workers from exposures. Chapter 10 discusses different types of personal protective equipment and their uses and limitations.

Occupational noise and ventilation are occupational environmental problems that have to be dealt with. Chapter 11 gives guidance about responsibilities and procedures to deal with these problems.

The last chapter, which discusses bloodborne pathogens, addresses the problems of infectious diseases and how to deal with them accordingly.

The supplements cover some elements that are included in the chapters, but are elements that are specific to certain organizations. These supplements include Welding, Materials Handling and General Housekeeping, Transportation Safety, Overhead Hoist and Slings, and Portable Tools and Machine Guarding.

Acknowledgments

I would like to give a profound thank you to Zachary S. LeMasters for his dedication in assisting me with this publication. The countless number of hours he spent researching the materials for this publication are to be commended. His efforts in editing some of the chapters are greatly appreciated.

Finally, I would like to thank Professor H. Ilkin Bilgesu, PhD, for his computer skills and reorganizing the chapters into a single coherent text. His efficiency and most especially, his patience are greatly appreciated.

Author

Daniel E. Della-Giustina is a Professor in the Department of Industrial and Management Systems Engineering and the Safety and Environmental Management Program, College of Engineering and Mineral Resources at West Virginia University, where he has been a faculty member for 25 years. Dr. Della-Giustina earned a PhD Degree in Safety, Health, and Higher Education from Michigan State University, and a Master of Arts and Bachelor of Arts degrees in Liberal Arts and Behavioral Sciences from American International College, Springfield, Massachusetts. He currently serves as a member of the Board of Trustees at American International College.

Dr. Della-Giustina, a professional member of American Society of Safety Engineers, has published more than 125 articles and 14 textbooks in the discipline of safety, health, and environmental studies. He has presented scholarly papers at numerous meetings and conferences at the national and international levels. He has presented papers at the International Sports Medicine Conference in Brisbane, Australia; the 2nd International Conference on Emergency Planning and Disaster Management, Lancaster, UK; Crime and Its Victims—International Research and Public Policy Issues, (NATO Conference) Tuscany, Italy; and Fitness and the Aging Driver, Stockholm, Sweden. He currently serves as editor of *The Safety Forum*, published by the School and Community Safety Society of America.

Dr. Della-Giustina is a former administrator of the American Society of Safety Engineers' Public Sector Division. At the 1995 ASSE Professional Development Conference, Orlando, Florida, he was presented with the ASSE Division's Safety Professional of the Year award, and in 2001, was presented the Safety Professional of the Year Award in West Virginia by Governor Cecil Underwood. He has appeared as an expert witness in safety and health liability cases for the past 20 years in numerous cities throughout the United States. He has served as consultant to various public, volunteer, and industrial fire brigades in the areas of disaster preparedness and emergency systems. During the past 25 years, Dr. Della-Giustina has been a member of numerous committees with the American Society for Testing and Materials (ASTM). In January 1998, he was appointed a principal member of the National Fire Protection's Technical Committee on Industrial Fire Brigades Professional Qualifications. This standard was adopted in 2002 by the NFPA. He also serves as a member of the Board of Hazard Control Management.

The American School and Community Safety Association has presented Dr. Della-Giustina with its Scholar Award (twice) and its Presidential Citation, as well. His distinguished record as a researcher, author, teacher, and administrator has made him a national leader in the safety profession.

In September 2002, Dr. Della-Giustina was inducted into the Safety and Health Hall of Fame International at the National Safety Council's Annual Convention in

Atlanta, Georgia. His most recent honor was in June 2005 when he was elected Fellow in the American Society of Safety Engineers. Both of these honors are the highest in each of these professional safety societies.

1 Introduction

SAFETY POLICY

The safety and health of all employees should be the first priority and value of any business. It should therefore be the policy of any business to provide a safe and healthful work environment for all of its employees. The business should have in place a continuous safety and health program to fulfill this goal.

With this in mind the safety rules contained within this book have been formulated to assist and protect you and your business. They were revised to keep up with the latest technological developments, changing conditions, and improvements in equipment and procedures as well as the requirements set forth by the Occupational Safety and Health Administration (OSHA).

Safety pays, and it is a good investment for all concerned. It eliminates suffering and lost wages. It also improves production, maintains efficacy, reduces waste, and generally provides a sense of well being for all employees and their families.

With these considerations, we need to continue to invest time and money toward continuous improvement of the safety and health environment. We expect employees to assist with these efforts by obeying the safety rules and procedures outlined in this safety manual. If you only protect one employee or family member from having to deal with the pain and suffering caused by a workplace incident then all the efforts in safety are worth it. However, you will protect more people than you could imagine in your time when working toward a safe and healthful work environment.

INTRODUCTION TO OSHA

The Occupational Safety and Health Act of 1970 was signed by President Nixon on December 29, 1970, and became effective on April 28, 1971. It requires every employer to provide a work environment that is free from recognized hazards that cause or are "likely to cause" death or serious physical harm. Each employer is required to comply with safety and health standards that are specifically spelled out by law.

The OSH Act authorized OSHA to regulate private employers in the 50 states, the District of Columbia, the Virgin Islands, American Samoa, Puerto Rico, Guam, and the Trust Territories of the Pacific Islands. According to figures from the Department of Labor, more than 89 million workers are covered by the Act. The statute also authorized the government to conduct research on occupational hazards and ways to correct them, and to oversee and approve state regulatory programs.

OSHA is an agency of the department of labor and is administered by the assistant Secretary of Labor. Three agencies administer the major requirements of the OSH Act: OSHA, the National Institute for Occupational Safety and Health (NIOSH), and the Occupational Safety and Health Review Commission (OSHRC).

OSHA's Responsibilities

- OSHA sets the health and safety standards that employers must meet to be considered in compliance with the act and it enforces those standards by inspecting workplaces in private industry and citing conditions that fail to meet the requirements.
- OSHA generally proposes monetary fines for violations that it deems to be serious. It also monitors the performance of state agencies that wish to set and enforce their own standards.
- OSHA requires employers to maintain records that may be used to judge compliance and to measure incidents involving work-related injuries and illnesses.
- OSHA may inspect any workplace under certain circumstances. It coordinates its regulatory activity with the activities of other federal agencies that regulate workplace safety and health in certain industries, by entering into agreements that delineate the responsibilities of the respective agencies and to avoid confusion and duplication of effort.
- OSHA furthers health and safety training for employers and employees by conducting training at a facility in Des Plaines, Illinois, and by funding courses though regional grants to other organizations. OSHA also uses the consultation branch to help give assistance to smaller employers who may not have the budget for a full-time safety staff.
- OSHA investigates complaints by employees who claim they have suffered retaliation from their employers for reporting safety hazards or from asking OSHA to inspect their workplace.

The act directs OSHA to develop and issue standards though a public rule-making process. Employers must comply with those standards as they would with any statutory requirement.

Record Keeping Requirements

Two kinds of records are required by OSHA. Section 8(c) of the OSH Act requires employers to maintain records necessary or appropriate for enforcing the Act. They are:

1. Affidavits, certificates, medical forms, measurements, and other such documentation that help the agency gauge the employer's compliance with specific safety and health standards.
2. Logs and supporting documents showing the number, nature, and circumstances of injuries and illnesses that occur in the workplace.

OTHER REGULATIONS AND AGENCIES

Fire Prevention Codes

These codes are concerned with safety regulations that relate to fire protection equipment, maintenance of building and premises, hazardous materials, processes, and machinery used in buildings. They are normally issued by authorities like Building Officials and Code Administration International, Inc. (BOCA) or the National Fire Protection Association (NFPA). It is important to check with the state fire marshal or other authorities to know which codes are applicable to your area.

Building Codes

Building codes address fire safety requirements with respect to the construction of buildings. They cover the basic structural requirements, structural integrity, and fire protection as related to structural elements, means of egress, interior finish, vertical and horizontal openings, and many other areas that are directly related to fire protection.

As with the fire codes it is important to know the codes applicable to your area. These codes are normally issued by the same agencies as the fire prevention codes.

NFPA 101, Code for Safety to Life from Fire in Buildings and Structures

The committee on Safety to Life of the NFPA initiated the Life Safety Code in 1913. The committee dedicated its efforts to the loss of life due to fire and its causes. Many pamphlets were published pertaining to standards for the construction and arrangement of exit factors, schools, etc., which form the basis of the present building code.

Today's codes include standards for all types of occupancies.

Company Policy

The employer establishes a policy that should be followed by all employees. These policies are normally above and beyond the standards and regulations required by OSHA and other regulatory agencies.

REVIEW QUESTIONS

1. The National Fire Protection Association (NFPA) initiated the ______________ in 1913?
2. What Section of the OSH Act requires employers to maintain records?
3. Name two of the three agencies that administer the major requirements of the OSH Act.
4. Who was the President who signed the 1970 Occupational Safety and Health Act?
5. Building Codes addresses fire safety requirements with respect to ______________?

BIBLIOGRAPHY

Kohn, James P., Friend, Mark A., and Winterberger, Celeste A. Fundamentals of occupational safety and health, Government Institutes, 1996.
Schneid, Thomas D. and Schuman, Michael S. *Legal liability*, Aspen Publishers, 1997.
Marcum, Everett C. Modern safety management practice, Worldwide Safety Institute, 1978.
Anton, Thomas J. *Occupational safety and health management*, Irwin-McGraw Hill, 1989.

2 Hazard Communication and Hazardous Materials Handling

INTRODUCTION

This chapter will introduce the material that will aid you in understanding what a hazard communication program should entail and what you need to know to comply with the OSHA Hazard Communication Standard 29 CFR 1910.1200. You will also learn how to properly warn employees about hazardous chemicals and substances in the workplace.

The first thing discussed in this chapter is a written program for hazard communication that is going to be important in staying in compliance with the requirements set forth by the OSHA standard. The specific methods described are used as a sample written program, and are for illustrative purposes and not intended to make up your entire program. Other effective methods can be substituted into this sample to satisfy the needs of the company for which you are working.

The Hazardous Materials Handling section will provide guidance in the safe storage, handling, and accounting for hazardous materials.

DEFINITIONS

CAS Number: The unique identification number assigned by the Chemical Abstract Service to specific chemical substances.

Combustible Liquid: Any liquid having a flashpoint at or above 100 degrees, but below 200 degrees.

Container: Any bag, barrel, bottle, box, can, cylinder, drum, reaction vessel, storage tank, tank truck, or the like that contains a hazardous substance.

Distributor: A business, other than a manufacturer or importer, that supplies hazardous substances to other distributors or to employers.

Emergency: Any potential occurrences such as, but not limited to, equipment failure, rupture of containers, or failure of control equipment that may or may not result in a release of a hazardous substance into the work environment.

Employee: Every person who is required or directed by any employer to engage in any employment, or to go to work or to be at any time in any place of employment.

Exposure or Exposed: Any situation arising from work operation where an employee may ingest, inhale, absorb through the skin or eyes, or in any way come in contact with a hazardous substance.

Hazard Warning: Any words, pictures, symbols, or combination thereof appearing on a label or other appropriate form of warning that convey the health hazards and physical hazards of the substance in the container.

Hazardous substance: Any substance that is a physical or a health hazard or included in the List of Hazardous Substances prepared by the Director Pursuant to Labor Code section 6382.

Health Hazard: A substance for which there is statistically significant evidence based on at least one study conducted in accordance with established scientific principles that acute or chronic health effects may occur in exposed employees.

Identity: Any chemical or common name that is indicated on the material safety data sheet (MSDS) for the substance. The identity used shall permit cross references to be made among the required list of hazardous substances, the label, and the MSDS.

Immediate Use: The hazardous substance will be under the control of and used only by the person who transfers it from a labeled container and only within the shift wherein it is transferred.

Label: Any written, printed, or graphical material displayed on or affixed to containers of hazardous substances.

Material Safety Data Sheet: Written or printed and even possibly computerized material concerning a hazardous substance that is prepared in accordance with section 5194(g).

NIOSH: The National Institute for Occupational Safety and Health.

Oxidizer: A substance other than a blasting agent or explosive as defined in section 5237(a), that initiates or promotes combustion in other materials, thereby causing fire, either of itself or through the release of oxygen or other gases.

Physical Hazard: A substance for which there is scientifically valid evidence that it is a combustible liquid, a compressed gas, explosive, flammable, an organic peroxide, an oxidizer, pyrophoric, unstable, or water reactive.

Responsible Party: Someone who can provide additional information on the hazardous substance and appropriate emergency procedures, if necessary.

Substance: Any element, chemical compound, or mixture of elements and/or compounds.

Workplace: Any place, and the premises appurtenant thereto, where employment is carried on, except a place of health and safety jurisdiction over which is vested by law in, and actively exercised by, any state or federal agency other than the division.

RESPONSIBILITIES/PROCEDURES

Hazard Communication

The overall coordination of the facility hazard communication program should be handled by the occupational safety and health manager (or other technically qualified designee). This coordinator should act as the senior facilitating official who is accountable for the entire program.

Employers shall provide employees with information and training on hazardous chemicals in their work area at the time of their initial assignment. The employee also needs to receive additional training when a new hazard is introduced into the work process or just as a refresher to once again make the employee aware of the dangers associated with handling hazardous chemicals in the work area.

The employer shall inform the employees about any operations in their work area where hazardous chemicals are present, along with training them on written hazard communication program and the location where it can be found in the plant. Not only is it a good idea to have a hard copy, an electronic copy is also beneficial because it makes the program more readily available to the employees and supervisors. The hazardous communication program will include a list of the hazardous chemicals and material safety data sheets. These two items are required to be in compliance with the OSHA regulations.

Training

Employee training shall include the methods and observations that may be used to detect the presence or release of a hazardous chemical in the work area. For example, monitoring conducted by the employer, continuous monitoring systems, visual appearance, or odor of hazardous chemicals being released.

The training shall include the physical and health hazards of the chemicals in the work area.

The training shall include the measures employees can take to protect themselves from these hazards, including specific procedures the employer has implemented to protect employees from exposure to hazardous chemicals, such as appropriate work practices, emergency procedures, and personal protective equipment to be used.

The employee shall also know the details of the hazard communication program developed by the employer, including an explanation of the labeling system and the material safety data sheet, and where employees can obtain and use the appropriate information.

Initial training must be conducted for all employees. New employees must be trained prior to their initial assignment.

Employees must receive additional training whenever:

- New hazardous substances are introduced into the workplace.
- Exposures to hazardous chemical change.
- Employees are subject to increased exposure due to changes in work practices, processes or equipment.
- Additional information about the hazardous substance in the workplace becomes available.

HAZARDOUS MATERIALS HANDLING

Chemical Inventory

A business should determine what hazardous materials will be used and maintain a current list of kinds and amounts. The location of these materials should also be noted to aid in an emergency situation. MSDSs should be obtained and kept on hand for reference as well as employee training, in accordance with the Hazard Communication Program.

Labeling and Storage

All hazardous substances must be stored and labeled properly. All containers should be compatible with the material they contain, no matter how minute the quantity, and properly labeled to prevent accidental misuse. The NFPA Standards and 29 CFR 1910 dictate proper storage, grounding, and dispensing of hazardous materials. Dangerous chemical reactions can result if certain substances are mixed together.

Chemicals such as paints, thinners, solvents, and fuels have specific storage requirements that should be carefully followed. Temperature, ventilation, vibration, and proximity to other substances can have adverse effects on certain hazardous materials.

Transferring of chemicals should be carefully done according to manufacturer's procedures with containers grounded to prevent fire or explosion from static electric charges. The NFPA has a label known as the "704" diamond, which informs users at a glance of the properties of a substance. The diamond consists of four different colors.

Personal Protective Equipment

All necessary engineering, site-specific and mandated personal protective equipment (PPE) must be used to prevent hazardous materials from harming personnel, property, and the environment. Records of issuance and training in the use and maintenance of PPE must be kept.

A program to train new employees and periodically update training of current employees should be in place. Any medical screening and records, such as a pulmonary function test for respirator wearers should be maintained, with deficiencies noted for correction.

The safety manager should make sure that each employee is using his or her PPE properly and not disabling or bypassing engineering safeguards. Any outside contractors who handle hazardous materials should be required to submit written safety procedures and be periodically monitored to ensure compliance.

Spill Plan

Any spillage of hazardous materials should be cleaned up immediately. Any damaged containers of hazardous substances should be reported, repaired, replaced, or otherwise contained immediately.

A viable spill plan to protect personnel, property, and the environment should be developed. Equipment for use in spill incidents should be kept on hand and properly maintained. Employees should be trained to use this equipment and some of those who have been trained should be present during each shift.

Periodic spill containment drills can be very helpful in quickly correcting a potentially dangerous and costly situation. Coordination and notification procedures with the local emergency services as well as state and federal environmental-protection officials should be instituted and maintained. They may want to know the type, amount, and location of any hazardous substance that your organization stores on site in order to be more effective in their response should they ever be called upon.

CHEMTREC

What Is CHEMTREC?

CHEMTREC stands for Chemical Transportation Emergency Center. It is a public service of the Manufacturing Chemist Association located at its offices in Washington, D.C.

CHEMTREC provides immediate advice for those at the scene of emergencies, then promptly contacts the shipper regarding the chemicals involved for more detailed assistance and appropriate follow-up. CHEMTREC operates around the clock—24 hours a day, 7 days a week—to receive direct-dial toll-free calls from any point in the continental United States through a wide-area telephone service number (800-424-9300).

Shippers, including MCA members and nonmembers, are notified through pre-established phone contacts, providing 24-hour accessibility via information operators or through cooperation of fire and police services.

As circumstances warrant, the National Transportation Safety Board or appropriate offices of other agencies may be notified.

CHEMTREC's capabilities have been recognized by the Department of Transportation, and a close and continuing relationship is maintained between CHEMTREC and the Department.

What It Is Not

Because chemicals find so many uses and have such a wide range of characteristics, there is much need for information about them, including their composition and purity, physical and chemical properties, effects on people and the environment, sources of supply, etc. It is important to understand that CHEMTREC is not intended and is not equipped to function as a general information source, but by design is confined to dealing with chemical transportation emergencies. CHEMTREC should not be called for problems other than chemical cargo emergencies.

Mode of Operation

CHEMTREC's number has been widely circulated in professional literature distributed to emergency service personnel, carriers, and the chemical industry, and has been further circulated in bulletins of governmental agencies, trade associations etc.

Shipping documents of participating companies are requested to include the following: "For help in chemical emergencies involving spill, leak, fire or exposure, call toll free 800-424-9300 day or night.

An emergency reported to CHEMTREC is received by the communicator on duty, who records details in writing and by tape recorder. The communicator then attempts to determine the essentials of the problem. This allows for him or her to provide information on the chemicals that are reported to be involved in the incident. This in turn gives specific indications of the hazards and what to do for that particular incident. Information for the chemicals is readily available as furnished by the producer. Trade names and synonyms of names are cross referenced for easy identification no matter what name is given.

The communicators are not scientists. They are chosen for their ability to remain calm in emergency situations. To preclude unfounded personal speculations regarding a reported emergency, they are under instructions to abide strictly by the information prepared by the technical experts for their use.

Having advised the caller, the communicator will then notify the shipper by phone. The known particulars of the emergency are relayed at this time. The responsibility for further guidance including dispatching personnel to the scene or whatever seems warranted passes to the shipper.

Identification of product and the shipper is important. Shipping papers are carried by truck drivers and in the engine compartment of trains. Car and truck numbers and carrier names can be helpful in tracking unknown cargoes. Although proceeding to the second stage of assistance becomes more difficult if the shipper is not known, the communicator is armed with other resources to fall back on.

Mutual aid programs exist for some products where one producer will service field emergencies involving another producer's goods. In such cases, initial referral may be in accord with the applicable mutual aid plan rather than direct to the shipper.

In Canada, the Canadian Chemical Procedures Association operates a Transportation Emergency Association Program (TEAP) through regional teams prepared to give phone and field response.

Many individual companies have well organized response capabilities for their own products, some of which preceded CHEMTREC by several years. This program does not seek to displace these, but rather collaborates with them and enhances their effectiveness. CHEMTREC's single telephone number affords this opportunity.

REVIEW QUESTIONS

1. A Hazardous substance is defined as ______________.
2. A written program is required to achieve ______________ with the requirements of the standard.
3. Periodic spill ______________ can be very helpful in quickly correcting a potentially ______________ and ______________ situation.
4. CHEMTREC stands for ______________?
5. The NFPA Standards and 29 CFR 1910 dictate proper ______________, ______________ and ______________ of hazardous materials.

TRUE OR FALSE

6. T or F—In Canada the Canadian Chemical Procedures Association operates a Transportation Emergency Association Program (TEAP) through regional teams.
7. T or F—CHEMTREC's number has not been widely circulated in professional literature distributed to emergency service personnel.
8. T or F—A program to train new employees and periodically update training of current employees on PPE should be in place.
9. T or F—A business should not determine what hazardous materials are used, but should try to maintain a current list of all chemicals for any industry.
10. T or F—Material Safety Data Sheets (MSDS) are produced by OSHA for the protection of Employers and Employees.

BIBLIOGRAPHY

Meyer, Eugene. *Chemistry of hazardous materials*, 2nd ed., Prentice Hall, 1989.

Hazard Communications Guide. J.J. Keller and Associates, Inc., 1988.

NFPA. Fire Protection Guide on Hazardous Materials, 9th ed., 1986.

NFPA 30. Flammable and Combustible Liquids Code Handbook, National Fire Prevention Association, 2006.

Material Safety Data Sheets, Genium Publishing Corporation, 1985.

OSHA Hazard Communication: a Compliance Kit, OSHA 3104, 1988.

OSHA Standard Code of Federal Regulation 1910.1200. U.S. Government Printing Office.
Della-Giustina Daniel E. *Safety and environmental management*, Van Nostrand Reinhold, 2006.

3 Job Safety Programs

INTRODUCTION

This chapter will discuss the safe work procedures needed to ensure that employees are properly trained on the safety requirements of their task and able to demonstrate this knowledge and performance in their every day duties. Finally, this is observed by management to provide encouragement, training, and corrective action as appropriate.

Four techniques will be discussed to help with these problems. They are Job Safety Analysis, Job Safety Training, Knowledge Reviews, and Work Observations.

DEFINITIONS

Accident: Any unplanned event that results in personal injury or property damage.

Hazard: Any condition that may result in the occurrence of, or contribution to an accident.

Job: A definite sequence of steps or separate activities that together lead to the completion of a work assignment.

Job Step: A single and separate activity that clearly advances a work assignment and is a logical portion of that assignment.

RESPONSIBILITIES/PROCEDURES

Job Safety Analysis

The Job Safety Analysis (JSA) is a process used to determine hazards of and safe procedures for each step of a job. A specific job or work assignment can be separated into a series of relatively simple steps. The hazards for each step can be identified and the solutions developed to control each hazard.

The JSA is a very important part of a safety program. The most effective way to accomplish these analyses is to involve all employees. Therefore, each worker, supervisor, and manager needs to be prepared to assist in the recognition, evaluation, and control of hazards through the JSA process.

Procedures

The JSA is a fairly simple process that can be as basic as four steps:

1. Selection of the job to be analyzed
2. Separating the job into basic steps
3. Identifying the hazards associated with each step
4. Controlling each hazard

Selecting Jobs

Selecting jobs to be analyzed is done on a priority basis. Select jobs based on the following guidelines.

1. Accident frequency
2. Accident severity
3. Judgment and experience (hazardous jobs)
4. Jobs with high turnover
5. New jobs
6. Non routine jobs
7. Routine jobs

Break the Job Down into Steps

The supervisor or safety official should first observe the job to be evaluated and then involve at least one worker who normally performs the job. Workers assisting in this process should understand the purpose of the JSA. Using the JSA worksheet, list the activities of a job that must be accomplished in their normal order of occurrence. Usually, three to four words are sufficient to describe each job step. The first word is generally an "action" or "do" word. Avoid making the mistake of making the job breakdown too detailed or too broad. If you have more than 15 steps, the job should be broken down into more than one JSA. Mutual agreement between the initiator of the JSA and the employee assisting should be accomplished.

Identify the Hazards - Once the basic steps of a job have been listed they can be carefully examined to identify potential hazards. Ask whether the employee can:

be struck by, strike against, be caught in; fall, trip, or slip: strain or overexert; or be exposed to health hazards. Are there machines, electrical, mechanical, or manual handling, dust, fumes, heat, gas, hand tools or other hazards inside or outside the work area. By identifying and understanding the hazards connected with each job, the entire job can be made safer and more efficient.

Developing Solutions

For every hazard associated with a job step, there should be a solution that offsets the hazard listed as a recommended safe job procedure or operation improvement/redesign. The following five questions should be asked:

1. Is there a less hazardous way to do the job (eliminate the hazard)?
2. Can the physical conditions that created the hazard be changed (tools, materials, and equipment that may not be right for the job)?
3. Can the job procedure be changed if engineering controls cannot be applied (i.e., performing work at the beginning of the shift instead of the end)?
4. Can the job be eliminated or the frequency of performing the job reduced (i.e., frequent maintenance repairs that could be prevented if the right, or more effective, replacement parts are used)?
5. Can PPE be used?

Key Points for Success

The main purpose of the JSA procedure is to prevent accidents by anticipating and eliminating hazards. Among their many other advantages, JSAs provide training to new employees on safety rules and how the rules are applied to the trainee's work. To make your JSA as successful as possible keep the following points in mind:

- Involve employees and let them know what is being done. Get their input regarding hazards, and possible solutions to those hazards, in either group or individual discussions.
- Review the completed JSAs with employees to ensure nothing has been omitted.
- Use the JSA form to simplify the process.
- Use JSAs to train new and transferred employees, or to provide refresher training.
- JSAs should be reviewed and updated periodically based on employee and supervisory input.

JOB SAFETY TRAINING AND KNOWLEDGE REVIEWS

As stated in the OSH Act, "Each employee shall comply with the occupational safety and health standards and all rules, regulations, and orders issued pursuant to this Act which are applicable to his/her own action and conduct." Furthermore, employers shall provide training in the various precautions and safe practices associated with the employee's job task and shall ensure that employees do not engage in the activities to which this chapter applies until such employees have received proper training in the various precautions and safe practices required by this chapter. Where training is required, it shall consist of on-the-job training or classroom-type training or a combination of both.

Sufficient training should be conducted to ensure that the employee is knowledgeable concerning the safety precautions associated with each job task. In addition to documenting that the initial training was conducted, the supervisor is responsible for ensuring that the employee both understands what he or she has learned and demonstrates this knowledge when performing the job task.

A form should be developed to be used company-wide to document all safety training. At a minimum, this form should consist of employees' name, Social Security number, kind of training performed, date of training, and signature of trainer. This documentation can be housed in the employee's personnel file or in a computerized system.

WORK OBSERVATIONS

To maintain an effective safety program the employer must perform periodic safety work observations to ensure that (1) the employee indeed has the knowledge to perform the work task as trained and (2) is indeed performing the work task as trained.

Management should conduct these work observations and document them on a form. The form should describe the employee's ability to perform the task correctly and what corrective action or retraining was needed. It is important that the management document these observations as a means of demonstrating to OSHA that they are monitoring the effectiveness of their safety program and ensuring employee compliance to established safety requirements. This information should be included with the personnel file or input into a computerized system.

An easy way to develop this work observation checklist is to take each individual work task and subdivide it into its safety components. An example of this is as follows:

Use of grinding wheel:

- Is eye protection used by employee?—yes/no
- Is grinder shield in place and used?—yes/no
- Any protective gloves being used?—yes/no
- Is grinder wheel in good condition?—yes/no
- Is employee using the proper methodology?—yes/no

If the supervisor were to answer no to any of these questions based on observation of an employee's work, he or she would interview the employee to determine the "root cause" of the deviation and to take corrective action. For example, if eye protection was not worn, the supervisor would determine:

- If the employee was trained on the use of proper eye protection.
- Did the employee have proper eye protection?

- Was proper eye protection defect free?
- Did the eye protection fit properly?
- Did employee choose not to wear proper eye protection?

Predicated on the answers to these questions, the supervisor can determine what corrective action is required to correct the deviation. The supervisor would document his or her actions on the form, state their corrective action and note any follow-up actions for future considerations, i.e., replaced defective eye protection, retrained employee on proper use, and made a notation to follow-up observation tomorrow afternoon.

SAFETY KNOWLEDGE REVIEW

Management should perform a JSA on all work tasks to determine employee compliance. This is the "Safety Knowledge Review" section of the program that is used to examine employees' safety knowledge associated with their job function. Using the grinding wheel as an example, the following questions could be asked when appropriate or on an annual basis:

Q. When using the grinding wheel what is the proper eye protection you should wear?
A. American National Standards Institute (ANSI)-approved dust proof goggles should be worn to protect my eyes when using the grinding wheel.
Q. How do you inspect the grinding wheel before use?
A. Check for the shield to be in place. Check the wheel for any defects, cracks, loose nuts, etc. Check the power cord for frayed wire, and damaged plug, etc.

Safety knowledge reviews and work observations should be routinely performed by the supervisor to measure the effectiveness of the company's safety program, training, PPE, compliance etc. At a minimum, the work observation form should include the employee's name, job title, year/date, supervisor name, work task(subject observed), OK/deviation, comments, and corrective action. This documentation should be maintained in the employee personnel file or a computerized file designed for this purpose. This information will be helpful in evaluating the company's accident prevention plan and will be used during an OSHA audit to document the company's safety training requirements.

REVIEW QUESTIONS

1. List the four basic steps of the JSA process.
2. ______________ is any condition that may result in the occurrence of, or contribute to an accident.
3. The ______________ is a process used to determine hazards of and safe procedures for each step of a job.

4. Each employee shall comply with the occupational safety and health standards and all ______________, ______________, and ______________ issued pursuant to this Act that are applicable to his/her own action and conduct.
5. What are the seven guidelines for job selection in the JSA process?

True or False

6. T or F—It is important to review the completed JSAs with employees to ensure nothing has been omitted.
7. T or F—Management should perform a JSA on all work tasks to determine employee compliance.
8. T or F—JSAs should never be reviewed or updated since nothing will ever change, and you don't need employee or supervisory input.

BIBLIOGRAPHY

Kohn, James P., Friend, Mark A., and Winterberger, Celeste A. Fundamentals of Occupational Safety and Health, Government Institutes, 1996.

Bird, Frank E., and Germain, George L. Practical Loss Control Leadership, International Loss Control Institute, 1996.

Job Hazard Analysis, OSHA 3071.

Germain, George L., Arnold, Robert M., Rowan, J. Richard, and Roane, J. R. Safety, Health and Environmental Management, AEI and Associates, 1997.

Della-Giustina, Daniel E. *Safety and environmental management*, Van Nostrand Reinhold, 2006.

4 Accident and Incident Investigation

INTRODUCTION

Many organizations utilize the term "incident" when referring to any unplanned event or event sequence. The reasoning behind their concept is the limitations the word "accident" connotes. When we discuss losses from incidents, safety professionals know they can take many different forms, i.e., damage to property, equipment, materials, and even the environment, as well as injury, illness, death, and disease. An accident represents some failure on the part of management to exercise control over conditions and actions in the workplace.

Planned accident investigation within a company or enterprise is used to identify four basic elements:

- Causative factors
- Trends
- Patterns for specific employees
- Establishing relevant facts in cases involving litigation

By utilizing accident investigation information in this manner future accidents can be prevented, unsafe work habits be corrected, and employee training needs can be identified and targeted.

Most occurrences of accidents are due to a failure of management to be able to control acts and conditions in the workplace. Once an accident occurs, management has the obligation to ensure that a similar act does not recur. This is achieved through proper accident investigation.

To be effective, accident investigation should have a plan to be executed by management and other trained individuals. These plans should be designed to satisfy corporate policies and legal issues. The plan should always produce an investigation to determine the cause of the accident and what measure must be taken to prevent it from happening again.

Responsibilities/Procedures

Management will be responsible for developing an accident investigation plan and providing the employee training required to perform their accountabilities under the plan. Employees are responsible for understanding their accountabilities under the plan and performing their required duties as appropriate.

Goal

The primary goal of conducting an accident investigation is to determine the "root cause" and implement corrective action to prevent the risk of future occurrence. Supervisors should be thorough in their investigations in order to achieve this goal. Depending on the seriousness of the accident, upper management, outside consultants, or OSHA may be involved in the investigative process.

Objectives

When establishing the company's accident investigation procedures and forms, all affected parts of the organization should be brought together to determine what information they need and why. This might include claims department, the safety department, legal department, workers compensation operations, etc. Once this has been completed, forms should be developed. "Occupational Injury and Illness" and "Motor Vehicle" are examples.

The following elements should be utilized by supervisors for conducting accident investigations:

- An established policy and procedure for conducting accident investigations, outlining the requirements and necessary steps for conducting thorough investigations.
- Supervisors should be familiar with the procedures and steps covered in the accident investigation to ensure that the corporation's goals are fully met through this process.

- The severity and scope of the accident will determine who needs to be involved in the investigative process, and what investigative tools are applied.
- The administrative requirements should be clear, concise, and to the point, i.e., what information needs to be captured, what time lines adhered to, etc.

THE ACCIDENT INVESTIGATIVE PLAN

There are a number of reasons for the investigations of accidents, one being to gather facts. Another is to take action and prevent similar situations from occurring again. This is the key to conducting a proper accident investigation. This should be prepared by corporate management in cooperation with the company's legal department. Once this plan is developed it should be initiated during the initial report of an accident's occurring. This plan will provide direction for the accident investigation.

The plan should consist of the following items:

1. Report of the accident. Once a report of a problem is received it is important that the appropriate medical services be dispatched, the emergency plan implemented, and the accident scene secured to prevent further injury and ensure that evidence needed for the investigation is still intact.
2. Notifications. The policy must spell out who needs to be notified in the case of an accident. The policy needs to dictate who will be involved in investigating certain events such as:
   - Fatalities
   - Injuries—serious and minor
   - Value—dollar and property loss
   - Motor Vehicles
   - Environmental—spill or release
   - Injuries to contract workers
   - Injuries due to intentional or negligent conduct
   - Product liability matters

   This notification policy should include some main personnel. The safety manager should be notified for investigative input, the human resources manager for worker compensation claims, the company physician, the legal department, and top-level management. In some cases, regulatory agencies such as OSHA, the Environmental Protect Agency (EPA) and the Department of Transportation (DOT) need to be notified.
3. Investigative responsibility. This is clearly a prerogative of upper management. Serious investigations need to be directed by a designated manager to ensure that all appropriate issues are addressed.

   In some cases a team with a manager responsible for the investigation will be formed. This is largely a function of a situation in which technical expertise is needed for the investigation

   In some cases, accidents are investigated by an outside source. This action can occur when no employees have the knowledge to perform the accident investigation. This would be common, for example, in a motor vehicle investigation in which an accident reconstruction is needed or in cases of a fire in which a fire investigator is needed to determine the origin.

4. Investigative methods. The accident investigative plan needs to provide guidelines to the investigator. If guidelines are not established before the accident, questions will arise that will delay the investigation and could result in destruction of important evidence. The accident investigation plan will provide guidance and techniques to be employed. Important areas that require a policy are:
   - Interviews—When interviews are conducted, who will be interviewed, and how will the interview be recorded? Will notes be used or a computer be used?
   - Isolating the scene—In addition to protecting others from a hazardous accident scene, the question that is common to ask is how long the scene can be secured for investigative purposes. The answer is, it depends on the circumstances surrounding the accident. The scene can be recorded by photographic equipment and if the accident is minor the isolation may be terminated. If the accident is major and requires experts, the investigation plan will dictate this time.

   A question that comes up in investigations is the use of videotaping. The use of videotaping is growing in popularity and has been employed in criminal investigations. There is no reason to not use videotaping to record accident scenes.

   Videotaping's advantages include providing a more realistic and graphic portrayal of the accident scene and the events leading up to the accident. Videotaping can also be used to record witnesses' testimonies. The use of such equipment is easy and cost effective in terms of recording evidence and reproducing copies.

Accident Investigation

Forms and record keeping procedures are based on the type of accident being investigated. A serious incident is considered an accident involving a fatality, the hospitalization of five or more workers as a result of a single event, and property damage of $100,000 or greater. These will require a more thorough investigation and must include the following information:

- Accident Investigation Report
- Fatality Report (Death Certificate, if applicable)
- Official Investigative Committee report
- Police Report (if applicable)
- Fire Department Report (if applicable)
- Coroner's Report (if applicable)
- Witness statements

- Photographs and sketches
- Relevant newspaper clippings

Incidents considered less than serious can be investigated by the immediate supervisor.

Supervisors

Supervisors might be accountable to conduct accident investigations at their organizational level. To facilitate this effort, management should train supervisors and ensure that they are familiar with the accident investigation plan. The following might be in a kit for the supervisors to use:

- High-visibility tape (to mark off area)
- Ruler and measuring tape
- Camera
- Pencil
- Graph and sketch paper
- Clip Board
- Containers and caps
- Plastic bags and envelopes
- Accident Investigation Report forms

Supervisors will have the responsibility for conducting accident investigations as outlined in their job description and dictated in the company's policy.

Serious incidents are to be reported immediately to the specific authorities outlined in the Accident Investigation Plan. The initial report should include the following information:

- Type of incident
- Date and time of incident
- Location
- Name and Social Security number of person(s) involved
- Job title
- Age
- Police report number
- Description and nature of the event
- Extent of the injury, illness or property damage
- Suspected causes and preliminary action taken
- Name of reporting individual
- Report of any governmental agency involvement

OSHA or DOT reporting should be conducted as outlined in the procedures and required by those agencies.

To improve effectiveness, supervisors and other accident investigation committee members should receive training adequate to perform the task for which they are accountable. The training should consist of the following elements:

- Introduction
- Purpose of accident investigation
- Incident reporting
- Incident response
- Evidence collection
- Causation analysis
- Corrective actions
- Accident Investigation Report
- Class exercises/application of principles

Based on the quality of investigating and reporting, the facts are frequently insufficient. The quality of accident investigation and reporting must be more sophisticated because this is a key element to the employer and the employees. A better approach and more efficient accident investigation program will contribute greatly to the reduction of future accidents. Once the analysis of the accident is reported, it leads to findings and conclusions, and that report must reach supervisors, managers, public officials, and others who can do something about the recommendations. Ultimately, these recommendations should lead to management and engineering changes, improved training, and changes in policies and procedures.

ACCIDENT INVESTIGATION FOLLOW-UP

Following are questions that could be asked in hopes of generating ideas for finding answers in the investigation and hopefully helping in the prevention of future accidents.

- Who experienced the accident?
- When did the accident occur?
- Where did the accident occur?
- What position was being worked?
- What task or job was being performed?
- What occurred?
- What were the causes of the accident?
- How can a recurrence be prevented?

REVIEW QUESTIONS

1. What are the four basic elements that a planned accident investigation can identify?
2. Serious incidents should be reported _______________ to the specific authorities outlined in the _______________ _______________ Plan.
3. A _______________ incident is considered an incident involving a fatality, or the hospitalization of five or more workers as a result of a single event.

4. The primary goal of conducting an accident investigation is to determine the ______________ ______________ and implement corrective action to prevent the risk of future occurrence.
5. What are seven items that the initial report should include?
6. What are some examples of questions that can be used to help generate ideas of how to prevent the incident from occurring in the future and discover how the incident occurred?

True or False

7. T or F—Supervisors are always accountable to conduct accident investigations at their organizational level.
8. T or F—Once an accident occurs management has the obligation to ensure that a similar act does not occur again.
9. T or F—The root cause is the primary goal of an accident investigation.
10. T or F—Serious incidents require a more thorough evaluation.

BIBLIOGRAPHY

Baker, Susan, O'Neill, Brian, Ginsburg, Marvin J., and Li, Guohua. *The injury fact book*, Oxford University Press, 1992.

Schneid, Thomas D. and Schuman, Michael S. *Legal liability,* Aspen Publishers, 1997.

National Safety Council 2008, *Injury Facts*, Itasca, IL. 2008.

Bird, Frank E. and Germain, George L. *Practical Loss Control Leadership*, International Loss Control Institute, 1996.

Reese, C.D. *Accident/incident prevention techniques*. New York, Taylor & Francis, 2001.

Root Cause Analysis Handbook: A Guide to Effective Incident Investigation. Rockville, MD, Government Institutes, 1999.

Della-Giustina, Daniel E. *Safety and environmental management,* Van Nostrand, Reinhold 2006.

5 Safety Committees

INTRODUCTION

One of the ways an organization can be successful in working on the elimination of workplace hazards is by instituting a safety committee. A well-structured and -trained safety committee will allow employees from all levels to participate, which will increase total safety awareness of all employees within the organization.

Normal functions of a safety committee are to:

- Audit and report, on a regular basis, unsafe conditions and practices, hazardous materials, and environmental factors within the organization.
- Review and update existing work practices and hazard controls.
- Make recommendations for improvements of existing safety and health rules, procedures, and regulations.
- Provide an effective way to communicate the safety and health information between and among all employees at every level within the organization.

RESPONSIBILITIES/PROCEDURES

The safety committee's role can include periodic site or area inspections, audits, assistance in accident investigations, and facilitation of employee hazard notifications.

Safety committee meetings are usually held monthly, at a minimum, to coordinate, integrate, assess, and implement safety activities.

Safety committee members are selected in a number of ways, such as appointed by members of management, election, volunteers, etc. Length of term may vary from 12 months to 2 years or more. The structure of any committee will be determined by the company philosophy; however, it is always beneficial when the employees have a more vital role in the committee. It is important to have a safety professional on the committee to act as a reference and provide direction. The committee must develop clear cut goals and expectations. Employee participation on these committees is discussed in more detail below, along with typical safety and health roles and responsibilities.

Neither federal nor state safety and health regulations require companies to form safety committees. However, they are widely recognized as an effective tool in generating employee involvement, raising employee awareness, and strengthening a company's overall safety and health program. One source that covers safety and health committees is contained in 29 CFR 1960, Federal Employee Safety and Health, Subpart F-Occupational Safety and Health Committees. Listed below are general provisions from safety and health committees required for Federal employees:

(a) The occupational safety and health committees described in this subpart are organized and maintained basically to monitor and assist an organization's safety and health program. These committees assist companies to maintain an open channel of communication between employees and management concerning safety and health matters in company workplaces. The committees provide a method by which employees can utilize their knowledge of workplace operations to assist agency management to improve policies, conditions, and practices.

(b) Companies may elect to establish safety and health committees meeting the minimum suggestions contained in this section. Committee organization for companies that elect to utilize the committee concept, safety and health committees, shall be formed at appropriate levels within the company.

(1) The principal function of the corporate level committee shall be to consult and provide policy advice on, and monitor the performance of the company-wide safety and health program.

(2) Committees at other appropriate levels shall be established at company establishments or groupings of establishments consistent with the mission, size, and organization of the company and its collective bargaining configuration. The company shall form committees at the lowest practicable local level. The principal function of the establishment (or local) committees is to monitor and assist in the execution of the company's safety and health policies and program at the workplaces within their authority. Committees shall have equal representation of management and non-management employees, who shall be members of record.

Management members of both national and establishment level committees shall be appointed in writing by the person empowered to make such appointments.

(3) Non-management members of establishment level committees shall represent all employees of the establishment and shall be determined according to the following rules:

(i) Where employees are represented under collective bargaining arrangements members shall be appointed from among those recommended by the exclusive bargaining representative:

(ii) Where employees are not represented under collective bargaining arrangements, members shall be determined through procedures devised by the company which provide for effective representation of all employees: and

(iii) Where some employees of an establishment are covered under collective bargaining arrangements and others are not, members shall be representative of both groups.

(4) Non-management members of national level committees shall be determined according to the following rules:

(i) Where employees are represented by organizations having exclusive recognition on an agency basis or by organizations having national consultation rights, some members shall be determined in accordance with the terms of collective bargaining agreements and some members shall be selected from those organizations having consultation rights, and

(ii) Where employees are not represented by organizations meeting the criteria of paragraph (i) of this section, members shall be determined through procedures devised by the agency that provide for effective representation of all employees. Committee members should serve overlapping terms. Such terms should be of at least two years duration, except when the committee is initially organized. The committee chairperson shall be nominated from among the committee's members and shall be elected by the committee members. Management and non-management members should alternate this position. Maximum service as chairperson should be two consecutive years. Committees shall establish a regular schedule of meetings and special meetings shall be held as necessary; establishment level committees shall meet at least quarterly and national committees shall meet at least annually. Adequate advance notice of committee meetings shall be furnished to employees and each meeting shall be conducted pursuant to a prepared agenda. Written minutes of each committee meeting shall be maintained and distributed to each committee member upon request and shall be made available to employees and to company management.

Employee Involvement: Rationale

An effective safety and health program cannot succeed without the active participation of all organizational elements. Ensuring worker safety and health is not just a management responsibility; employee involvement is the key to success of all safety and health activities. Because employees may have daily or direct contact with hazardous chemicals and potentially hazardous situations in their work environment, their ability to identify hazards and suggest or initiate corrective actions should be fully exploited as a means of increasing participation. Managers, employees, and union points of contact should use this intimate knowledge of the worksite to develop or augment an effective safety and health program.

Employee involvement and participation in the structure and operation of safety and health programs will foster increased and potentially lasting awareness, responsibility, and support for safety and health concerns and solutions to those concerns.

Employee Involvement: Element 1

For employees to become involved with their safety and health program, they must know about it and understand how it works for them. Several components of employee safety and health programs require direct, committed employee participation if the program is to succeed in helping to provide for a safe and healthful work environment.

Therefore, if you do not know of your worksite's employee safety and health program or how to participate in its application and operation, inquire of your supervisor.

Employee Involvement: Element 2

Communication among and between all worksite employees including management is essential for disseminating safety and health information. Employees must have access to effective modes of communication in order to express their feelings, concerns, and constructive ideas about safety and health on the job. Open lines of communication are essential for providing feedback related to corrective actions or progress made in implementing corrective actions.

Communication modes can be direct or indirect. Direct modes of communication affecting standard operating procedures and other procedures may include "open-door" policies, one-on-one discussions with immediate supervisors, open meetings, participation on related committees and work teams, participation in worksite inspections and accident/incident investigations, developing and designing safety and health procedures, reviewing and presenting safety and health training materials.

Modes of indirect communication may involve safety and health suggestion boxes and electronic mail systems. Unrestricted and functional communication systems can therefore significantly contribute to the promotion of greater employee input, feedback, and education about worksite mechanisms, policies and procedures, and employee safety and health protection by incorporating a flow of information aimed at correcting hazards and identifying the concerns of everyone.

Employee Involvement: Element 3

Open communication systems however, do not necessarily ensure that employee safety and health concerns will be addressed. Management must listen to and be held accountable for addressing all employee safety and health-related concerns and suggestions by acknowledging such concerns and responding to them in a timely way. Management indifference can discourage employee participation and potentially threaten employee safety and health. Prompt action applied appropriately and consistently is the best demonstration of management commitment to addressing safety and health concerns and fostering employee participation. The net result will be a shift in worksite behavior and practices toward an open, active, participative system for safety and health.

Employee Involvement: Options

Employees can become directly involved in maintaining worksite safety and health through several different types of activities that may either already exist or can be started as part of an employee safety and health program. These activities include:

- Union–management committee participation
- Specific-function committees
- Quality circles
- Site inspection participation
- Routine hazard analysis assistance
- Site safety requirements and standard operating procedures development or revision
- Training
- Participation in accident and incident investigations

Committee members need to have a broad understanding of the different roles, rights, and responsibilities traditionally attributed or assigned to upper management, managers, and supervisors, safety/loss control personnel and line workers. The following is a listing of typical rights and responsibilities assigned to various employee groups.

Rights and Responsibilities

All employees should be empowered to and are responsible for maintaining and promoting safety and health within their worksites. However, different, more specific, responsibilities can exist for different roles that individuals may have within an employee safety and health program. Specific responsibilities for safety professionals, managers/supervisors, employees, and employee representatives include:

Safety Professionals

- Administer an employee of a safety and health program to abide by their responsibilities within the workplace.
- Investigate and document employee complaints and reports of unsafe or unhealthful working conditions.
- Conduct required or necessary workplace inspections prior to implementing or requesting corrective actions.
- Ensure compliance with laws in regulations.
- Encourage employee participation and involvement.
- Perform record keeping.

Managers/Supervisors

- Serve as a primary source for employees to channel or voice employee concerns.
- Inform employees of appropriate channels for reporting concerns.
- Identify, report, and monitor actual or potential workplace hazards and concerns.

- Ensure abatement of hazards and monitor progress.
- Notify safety and health professionals of reported concerns.
- Ensure safety and health training and orientation for new employees.
- Ensure compliance with OSHA and company policies in the work areas.
- Ensure that proper safety equipment is available.
- Make sure of record keeping for safety meetings.

Employees

- Learn, understand, and comply with OSHA requirements and company safety and health policies.
- Review and understand safety and health educational materials posted or distributed in the work area.
- Become a proactive, safety and health conscious employee
- Assume greater personal responsibility for preventing or reporting safety and health mishaps.
- Fully understand and follow instructions based on job safety, analysis tasks (JSAs)
- Be provided access to and training on personal protective equipment.
- Do not commence work if unresolved questions or concerns exist related to any safety and health aspects of a job.
- Offer safety and health suggestions
- Support supervisors and safety professionals or other employees in their assigned roles in the safety and health program.
- Know and understand emergency plans and procedures.
- Immediately report workplace safety and health hazards.

Employee Representatives

- Represent designated employees in any and all safety and health matters.
- Provide information to employees concerning their right to a safe and healthful workplace.
- Notify safety professional of any observed or reported safety and health concerns and violations.

REPORTING UNSAFE CONDITIONS

Employees play a key role in identifying, controlling, and reporting hazards that may occur or already exist in the workplace though job-related activities performed at the worksite. Daily hands-on tasks performed routinely provide an intimate knowledge of safety and health conditions specific to each employee's work environment. Given this unique perspective, employees must have a reliable system for reporting unsafe or unhealthful working conditions as they are encountered on the job. A reliable reporting system is therefore integral to an effective safety and health program.

Safety and Health Hazard Reporting

Several systems for reporting safety and health hazards and concerns exist for employees that can be implemented for use at your facility. Examples of more common systems include oral reports to supervisors and union representatives, suggestion programs, maintenance work orders, and written forms that provide employee anonymity. Many worksites use a combination of part or all of these systems.

Oral Reports

Employee oral reporting can be one of the most effective and efficient methods for quickly relaying safety and health hazards and concerns to supervisors. If a supervisor is given an oral report, he or she is first responsible for assessing the report and determining whether or not he or she can correct the problem. He or she remains responsible for ensuring that the report is further communicated to the proper individuals within a pre-established safety and health concerns reporting system for expedient resolution. The supervisor is additionally responsible for providing timely feedback regarding decisions about corrective actions or any other information that is relative to the employee who initiated the oral report.

Although supervisors should encourage the use of oral reporting, this system will not guarantee safety and health concerns and hazard tracking or follow up, nor will it identify trends or patterns.

Suggestion Programs

The most frequently used type of written reporting system is a suggestion program that empowers employees to provide written safety and health suggestions for review and potential incorporation into an employee safety and health program.

Suggestion programs encourage proactive employee participation by providing an opportunity for employees to suggest creative approaches to maintaining worksite safety and health programs and reporting hazards and violations. If this system is used, supervisors must ensure that suggestions are collected, read, and responded to expediently and efficiently.

A "Stop" Program

Some private-sector worksites such as the Dupont Corporation have successfully implemented an innovative written reporting system known as the "Stop" program, which has as its cornerstone basic hazard-recognition skills. A similar system may work well in your organization.

Basically, through training, employees become adept at safety and health hazards recognition and empowered and encouraged to observe worksite conditions and report violations of other unsafe situations. Employees then record their observations on cards called "stop" cards, which are collected and given to the safety department for further follow-up, tracking, and corrective action.

As an added incentive for safety and health employee vigilance, some companies offer awards to employees who turn in the most cards with valid concerns over a specific period of time. Other companies impose quotas for the number of cards turned in. The extent to which this program succeeds depends largely on the safety and health culture espoused by different organizations , often reflected in the quality of safety and health training supported for employees.

Maintenance Work Orders

For unsafe conditions in the workplace, normally maintenance personnel must make the proper correction. Contact your appropriate maintenance dispatcher for specific instructions and procedures.

Employees should use this system only if a special high priority code for maintenance work orders involving safety and health hazards and violations exist. This will enable the maintenance supervisor to give these types of work orders higher priority than normal maintenance. Copies of these high-priority orders should also be delivered to the safety department so that ensuing corrective action can be tracked.

As a sole means of reporting, the work-order system is not sufficient. Although it can lead to the correction of hazardous conditions , it will not necessarily eliminate hazardous practices or provide for more imaginative approaches to standard conditions and procedures to improve worksite safety and health. Issuing work orders prior to determining the cause of employee complaints may eliminate hazardous conditions; however, the root cause of hazards may not get identified, and the hazardous condition may reoccur.

Written Forms

While some of the systems described above involve written forms, safety and health employee program managers and safety professionals or other appropriate individuals should feel free to design their own hazard reporting forms tailored to their worksite and employees. The flexibility in design can allow for voluntary anonymity by making employee self-identification optional. A box for the forms should be installed along with a specific location for the forms.

Reporting Unsafe Working Conditions

Remember that flexibility can be exercised in choosing or designing a reporting system that will work effectively for individual worksites. A policy should exist within your worksite that explicitly encourages employees to report all hazards and other safety and health concerns.

Improving a hazard prevention or employee program for safety and health is dependent upon the proper use of the information provided through various sources and channels that are available for communicating concerns. All employees should contribute to ensuring that the reporting policy is known and understood by everyone. The response should be swift and appropriate and followed up by a tracking system that assures implementation of any corrective action. The seeking of positive input from concerned employees also requires a benign environment free from any threat of harassment or reprisal.

REVIEW QUESTIONS

True or False

1. T or F—In order for employees to become involved with their safety and health program, they must know about it and understand how it works for them.
2. T or F—Safety and health professionals should feel free to make their own hazard forms tailored to their job site.
3. T or F—Federal safety and health regulations require companies to form safety committees.
4. T or F—Open communication systems ensure that employee safety and health concerns will be addressed.
5. T or F—Managers serve as a primary source for employees to channel or voice employee concerns.

BIBLIOGRAPHY

Anton, Thomas J. *Occupational safety and health management*, Irwin-McGraw Hill, 1989.
Lewis, Darcy. Safety committees, *Safety & Health* April, 1999.
Denton, Keith D. *Safety management improving performance.* McGraw-Hill, 1982.

6 Fire Loss Control Programs

INTRODUCTION

The purpose of a fire prevention plan or a fire loss control program is the prevention of fires and minimization of fire loss. The following elements contained in the Responsibilities/ Procedures are considerations of what such programs might contain. These elements include: Inspections, Education and Training, Fire Supression, Evaluation of Fire Possibility, Fire Prevention, Reports and Record Keeping, and

Communications. These elements should have complete support of upper management based on the fact that this is necessary for the development of implementation of loss control safety programs. All personnel must participate in a fire safety management program.

The success of any organization depends on the soundness of its management system; this is especially true in the management of fire safety. Certain techniques have been the hallmark of efficiency and profitability in the operation of any organization in order to utilize the successful management of a fire safety program. A basic knowledge of the available resources and fire safety organizations is essential. A basic premise of knowing where to go, who to contact, what facilities and equipment they possess, and their ability to respond will be of great assistance in organizing a plan of action.

RESPONSIBILITIES AND PROCEDURES

Hazard Inventory

All potential fire hazards should be identified. Hazards should be eliminated through modification, an engineering process, or possible substitution of a process or material that reduces the degree of risk.

Factors considered should include:

- All physical hazards concerned with the facility and its construction
- The procedures or processes used to produce the products or services
- The materials used in every aspect of the business
- The tools, equipment, and forms of energy used in the work procedures
- The likelihood of the workplace environment's changing in the future

Written Fire Plan

Your fire plan may contain the following:

- A list of the major workplace fire hazards and their proper handling, storage procedures, potential ignition sources. and procedures for their control, and the type of fire protection equipment or systems to be used for control.
- Names or regular job titles of people responsible for maintenance of equipment and systems installed to prevent or control fires.
- Names or regular job titles of people responsible for control of fuel source hazards.
- Written housekeeping procedures that will enable the employer to control accumulations of flammable and combustible waste materials and residues so they do not contribute to a fire emergency.
- Training requirements that ensure that employees receive training in four defined areas:
    1. Awareness of the fire hazards of the materials and processes to which they are exposed

    2. Knowledge of those parts of the fire prevention plan needed to protect themselves in case of emergency
    3. Review of the plan whenever it is changed, at least annually
    4. Training in the use of fire extinguishers provided in the workplace
- Procedures for regular maintenance of installed firefighting equipment, items that might be sources of ignition, and all fire extinguishers located in the facility.
- Proper marking and control of exits to prevent their blockage at any time. As part of the training of employees, each one must physically walk the route of evacuation so there is no misunderstanding about it in emergency situations.
- Proper identification and marking of emergency cut-off values. Persons designated to perform shut-down operations must be clearly identified and properly trained in their emergency roles.

Inspection Program

An inspection program should be initiated in addition to the housekeeping and critical parts inspections. Separate fire inspections and target special hazards specifically. A checklist can help guide the inspection and ensures inspection efficiency. Fire inspections may be done quarterly, monthly, or daily depending on the hazards involved. The inspection should begin with a thorough understanding of the layout of each building and each process utilized in the operation. Questions as to the degree of authority the inspector has should be settled before the inspection begins. The inspector must have access to every space or area, regardless who is in charge of these areas. Check outside, around the structure and roof area, and each floor down to the basement. If the individual responsible for fire protection in that building is available, he or she should accompany the inspector. Supervisors should be trained in the specifics of what to look at and look for in their areas of responsibility. See Appendix A for possible inspection items.

Regular Fire Drills

Fire drills verify the organization's readiness to handle a fire emergency effectively. The entire facility should be involved, using unannounced drills. The drills may be done by department or section. Management members should have specific responsibilities and be required to carry them out during an emergency as well as during the drill. Evacuation routes and exits should be posted in each work location.

Fire Responsibility and Accountability

All levels of management must be involved in the loss control program. The frontline supervisor is the key to a successful program. Supervisors have the most control over the variables that affect both the prevention and the proper control of emergencies. Supervisory control extends to:

- The primary causes of fires
- Fuel availability for fires to start
- Early reporting of fires
- Accessibility of the fire areas
- Evacuation of employees via specific predetermined routes based on the plan
- Resources for fighting incipient fires

Supervisors have broad knowledge of the products, materials, machines, equipment, processes, buildings, storage, and day-to-day hazards of the work. They should be given the responsibility for all the fire hazards and equipment in their areas.

Each building or area should be assigned to a specific person who clearly understands that he or she is responsible for the building and accountable for all aspects of the fire plan. This includes fire inspections and items on the inspection checklist, tools, and equipment being used by the employees.

The way to achieve responsibility and accountability is that nothing be left to chance or to assumptions. Everything must be clearly spelled out in detail to every manager and employee.

Management Involvement

Top management must be active in setting policy, establishing procedures, approving the written plan, and actively participating in all fire drills and inspections. Management must also insist that fire inspection results be reported regularly in management meetings to emphasize the importance of fire prevention to the whole management team. Executives must be more than just concerned; they must be committed and involved for effective fire loss control.

Training

Training that is conducted in addition to the training listed in the written fire plan must be based on prepared lesson plans. New employees must be trained and all employees must have an annual update.

Supervisors should be trained in the nature of fire and how it applies to their own environment. They should be trained in the specific hazards of the workplace. A permit system should be in place for controlling fires to ensure appropriate inspection.

Employee training should include:

- Fire potentials in their work procedures
- Department and plant rules which relate to fire loss control
- Emergency preparedness
- First aid training for burns

Fire Brigades

Fire brigades are an essential part of a plant emergency response organization. A fire brigade may be large and sophisticated or a small group with limited capabilities. In either case, the brigade must be equipped and know what its role is in the event of a situation. In the event of a fire there are six options that a employee fire brigade can use. They are:

Option 1: Employees will evacuate the building, no portable fire protection equipment available. Employer must provide a written emergency action plan, fire prevention plan, evacuation plan, and shutdown training.

Option 2: Employees will evacuate the building, portable fire protection equipment provided. Employer must meet same requirements as option 1, along with maintaining and testing portable fire equipment in compliance with standards.

Option 3: All employees will use portable fire equipment in their immediate work areas. Employer must provide training to all employees when first hired and annually thereafter on fire extinguisher use and selection.

Option 4: Portable fire equipment will be used only by designated employees in their assigned areas. Employer must provide training to designated employees as in option 3 and also an emergency action plan, evacuation training, action plan, and shutdown training.

Option 5: Portable fire equipment will be used by a fire brigade to fight incipient-stage fires only. Employer must provide training, a fire brigade organization statement, annual training, specific hazards training, and higher level training for instructors and leaders.

Option 6: The fire brigade will fight all fires including interior structural fires. Employer must provide a brigade policy and organization statement, training when assigned, training for special hazards, higher level training for instructors and leaders, brigade training, quarterly physical examinations of all members, and OSHA-required protective equipment and clothing.

Fire Brigade's Responsibilities

- Supervise department fire drills and exercises when a member is on vacation.
- Provide emergency scene first aid, CPR and AED if necessary.
- Operate firefighting equipment (e.g. ladders, hoses, extinguishers).
- Conduct inspections of particular departments.
- Implement emergency shutdown procedures.

The NFPA 600 standard covers the minimum requirements for organizing, operating, training, and equipping industrial fire brigades.

Cutting and Welding

Almost 6% of industrial fires are caused by portable cutting and welding equipment. The ignition sources are the sparks, hot slag or gas flame, and hot work pieces. Anything that is flammable or combustible is susceptible to fire ignition by cutting and welding. The most common are floors, partitions, roofs, wood, paper, plastics, liquids, and gases. Responsibility for cutting and welding rests with the cutter or welder, supervisor, and management for the safe use of welding equipment.

Management must establish procedures for approving work, designate an individual who is responsible for the authorization of cutting and welding, insist on proper training, select qualified contractors, and advise contractors about hazardous conditions involved when cutting and welding.

The supervisor is responsible for the safe handling of the equipment. He or she determines the hazardous conditions, protects combustibles from ignition by moving them to a safe location, sets the schedule and secures authorization from management, determines if a fire watch is needed if no fire watch the supervisor will conduct a final inspection after the work is done to detect or extinguish any fires.

Automatic Sprinkler Systems

There are several types of automatic sprinkler systems. These systems have piping of different sizes with sprinkler heads attached and connected to a large water supply. These heads are heat activated by a fire. They will then discharge water over the fire area. There are basically six types of automatic sprinkler systems.

1. The wet-pipe system. The pipes always contain water. Therefore, building temperatures must remain above freezing or provisions must be made for antifreezing solutions. The water does not flow until a sprinkler head starts discharging water. An alarm may signal when this happens. This type of system accounts for 75% of all sprinkler systems presently in operation.
2. The dry-pipe system. Water is not kept in the pipes. Thus, this system is used in buildings that have temperatures that drop below freezing. In this system, air or nitrogen is held in the pipes. The valve that controls the water flow must be installed in a heated area to prevent water in the valve from freezing.
3. The deluge system. This system permits all sprinkler heads to open and discharge water at the same time. A detection system in the protected area allows for water to flow through the pipes. This a good system for areas of storage for highly flammable materials.
4. The pre-action system. This is similar to the deluge system, except that the sprinkler heads are kept sealed for some time before the water comes through. This delay enables firefighters to put out the fire before it is covered by water.
5. The dry-pipe and pre-action systems. These are available for large structures where piping would be needed in unheated areas. More than one dry-pipe system would be needed. When the detection system detects a fire, an alarm sounds and water is fed into the pipes.
6. The recycling system. This is similar to the wet-pipe system with special heads, but the system automatically turns off the water when the fire is out. If the fire flares up again, the heads will discharge water again. This system is useful in protecting valuables from excessive water damage.

Foam Extinguishing Agents

Two main types of foam are in use today. These two types are chemical and mechanical; they differ in the way they are produced. Chemical foams are produced by a chemical reaction. Mechanical foams are produced by mixing a foam concentrate with water.

Chemical foam used to be made by pumping different chemicals into a mixing chamber over the object the foam was going to be used on (a tank for chemical storage for example). Chemical foam systems have functioned through the use of either single or double powder generators. The A powder aluminum sulfate in solution with B powder bicarbonate of soda and water forms a blanket on the surface if it is a flammable liquid, and from the oxygen, separates the vapors necessary to have combustion.

In recent years, mechanical foam has come to be used more frequently. The foam is made of hydrolyzed animal or vegetable protein, stabilizers, solvents, and an industrial germicide. The regular protein-type foams are proportioned into water at rates of either 3% or 6% and are suitable for ordinary hydrocarbon liquids. These protein foams are biodegradable, nontoxic, noncorrosive, and do not present any major clean-up problems.

The four ways to produce air foam are nozzle aspirating systems, in-line foam pump systems, in-line aspirating systems, and in-line compressed air systems. The names indicate where and how the air is injected into the water-concentrate solution to produce air foam.

Dry Chemical Extinguishing Systems

Dry chemical agents are effective on flammable liquids, greases and other class "B" fires. They are also effective on small fires in electrical equipment such as panel boards, switchboards, and other class "C" fires.

Dry chemical agents do not work well on deep-seated fires in ordinary combustible materials such as wood, paper textiles, and other class "A" fires. This is especially true where the cooling effect of water is needed for complete extinguishing of the fire.

Most dry chemical systems use sodium bicarbonate. This dry chemical agent works best to extinguish flammable liquids, but also works on electrical fires. One problem with this agent is that it is corrosive and can damage the finely polished metal surfaces found in electrical equipment.

Monoammonium phosphate is an agent used in multipurpose extinguishers ("ABC" type extinguishers). This is a especially desirable for the untrained person. These extinguishers are not to be used on fire involving combustible metals.

Portable Fire Extinguishers

Portable fire extinguishers are the most common type of private fire protection devices. These extinguishers contain a limited amount of extinguishing material and are to be used to put out smaller fires. A company needs to provide training to all employees who have access to an extinguisher. This would include how to operate the extinguisher and the right classification of extinguisher to be used for different fires. These extinguishers need to be maintained and inspected routinely to make sure they are charged and in good working order.

Classification of Extinguishers

- Class A extinguisher. For use on ordinary combustibles such as wood or paper. These extinguishers are normally water. They normally operate by changing water to steam, which has a cooling effect on the fires.
- Class B extinguisher. For use on flammable materials or combustible materials such as kerosene, gasoline, and grease.
- Class C extinguisher. For use on electrical fires. Utilizes a nonconducting extinguishing agent such as carbon dioxide or certain dry chemicals.
- Class D extinguisher. For use on combustible metals such as magnesium, titanium, sodium, potassium, and zirconium.

Fire Triangle

Three factors that create a fire are oxygen, fuel, and heat. These form a chemical reaction that starts a fire.

1. Oxygen. The atmosphere contains 21% oxygen by volume. During combustion, the oxygen necessary for oxidation comes from the surrounding air. When the oxygen content of the air falls below 16%, free burning of the fire will be curtailed.
2. Fuel. Fuel is a combustible substance that can be classified into one of the three physical states: solids, liquids, and gases.
3. Heat. Heat is the third component of fire. Heat sources from fire come from sparks, electrical, friction, etc.

Fire Tetrahedron

The concept of preventing the variables of the fire triangle from coming into contact with each other to initiate a fire is fire prevention. When a fire starts, it requires four variables to sustain combustion reaction. They are: fuel, oxygen, heat, and chemical chain reactions. These four variables represent the fire tetrahedron. Chemical chain reactions are a product of the combustion process.

APPENDIX A

While conducting an inspection, look for:

Brooms, pipes, and other debris stacked around or against circuit breakers
Containers of paint or solvent left unsealed and next to heat sources
Collections of flammable debris that could become ignited or supply fuel to a fire starting from some other source
Loose, frayed, or temporary wiring that could serve as an ignition point for fire
Non-explosion-proof lighting, fixtures, switches, etc. in areas where flammable vapors or dusts could accumulate and ignite
Fire extinguishers that are inoperable or cannot be reached because of materials blocking access
Unmarked or blocked fire exits or means of escape

Exit doors that open inward, so that passage from the area is more difficult (Life Safety Code)
Combustible materials stored near flame or spark-producing operations (grinding, gas cutting, or welding)
Fuel and oxygen lines that are unmarked or improperly identified
Absence of "No Smoking" signs in areas where flammable liquids, vapors, gases, or other highly combustible materials are stored or used; spark-producing tools in such areas
Inadequate ventilation in areas where painting, solvent cleaning, or other operations that produce flammable vapors or gases are performed
Fire extinguishers that are located in areas where they are not readily visible
Fire extinguishers that are not appropriate to the type of hazard in the area
Fittings that could allow pneumatic tools to be inadvertently connected to fuel or oxygen lines
Flammable liquids in unmarked or improperly identified containers

APPENDIX B

29 CFR 1910.38(b)

Fire Prevention Plan

(1) Scope and Application. This paragraph (b) applies to all fire prevention plans required by a particular OSHA standard. The fire prevention plan shall be in writing, except as provided in the last sentence of paragraph (1)(4)(ii) of this section.
(2) Elements. The following elements, at a minimum, shall be included in the fire prevention plan:
  (i) A list of the major workplace fire hazards and their proper handling and storage procedures, potential ignition sources (such as welding, smoking, and others) and their control procedures, and the type of fire protection equipment or systems that can control a fire involving them;
  (ii) Names or regular job titles of those personnel responsible for maintenance of equipment and systems installed to prevent or control ignitions or fires; and
  (iii) Names or regular job titles of those personnel responsible for control of fuel source hazards.
(3) Housekeeping. The employer shall control accumulations of flammable and combustible waste materials and residues so that they do not contribute to a fire emergency. The housekeeping procedures shall be included in the written fire prevention plan.
(4) Training.
  (i) The employer shall apprise employees of the fire hazards of the materials and processes to which they are exposed.
  (ii) The employer shall review with each employee upon initial assignment those parts of the fire prevention plan that the employee must know to

protect the employee in the event of an emergency. The written plan shall be kept in the workplace and made available for employee review. For those employers with 10 or fewer employees, the plan may be communicated orally to employees and employer need not maintain a written plan.

(5) Maintenance. The employer shall regularly and properly maintain, according to established procedures, equipment and systems installed on heat-producing equipment to prevent accidental ignition of combustible materials. The maintenance procedures shall be included in the written fire prevention plan.

1910.39 Sources of Standards. (OSHA)

The entire subpart is promulgated from NFPA 101-1970, Life Safety Code.

APPENDIX C

Appendix to Subpart E—Means of Egress

1910.38 Employee Emergency Plans

4. Fire prevention housekeeping. The standard calls for the control of accumulations of flammable and combustible waste materials. It is the intent of this standard to assure that hazardous accumulations of combustible waste materials are controlled so that a fast developing fire, rapid spread of toxic smoke, or an explosion will not occur. This does not necessarily mean that each room has to be swept each day. Employers and employees should be aware of the hazardous properties of materials in their workplaces, and the degree of hazard each poses. Certainly oil-soaked rags have to be treated differently from general paper trash in office areas. However, large accumulations of waste paper or corrugated boxes, etc., can pose a significant fire hazard. Accumulations of materials that are easily ignited and that can cause large fires or generate dense smoke, or may start from spontaneous combustion, are the types of materials with which this standard is concerned. Such combustible materials may be easily ignited by matches, welder's sparks, cigarettes, and similar low-level energy ignition sources.
5. Maintenance of equipment under the fire prevention plan. Certain equipment is often installed in workplaces to control heat sources or to detect fuel leaks. An example is a temperature limit switch often found on deep-fat food fryers found in restaurants. There may be similar switches for high-temperature diptanks, or flame failure and flashback arrester devices on furnaces and similar heat-producing equipment. If these devices are not properly maintained or if they become inoperative, a definite fire hazard exists. Again, employees and supervisors should be aware of the specific type of control devices on equipment involved with combustible materials in the workplace and should make sure, through periodic inspection or

testing, that these controls are operable. Manufacturers' recommendations should be followed to assure proper maintenance procedures.

APPENDIX D

NFPA 101, Code for Safety to Life from Fire in Buildings and Structures, 1991 Edition:
Chapter 2 Fundamental Requirements
Chapter 5 Means of Egress
Chapter 28 Industrial Occupancies
Chapter 31 Operating Features

REVIEW QUESTIONS

1. ______________ is the temperature at which a material gives off volatile vapors.
2. List the three types of fire detectors.
3. Class D extinguishers are suitable for fires with combustible ______________.
4. All employees will use ______________ fire equipment in their immediate work areas.
5. List three factors that make up the fire triangle.

True or False

6. T or F—For a fire brigade program to succeed, it must be recognized by management.
7. T or F—Standards for fire protection are established by the National Safety Council.
8. T or F—There are three classes of automatic fire detection systems.
9. T or F—Insurance carriers pay little attention to sprinkler system reliability.
10. T or F—Fuel is any combustible material—solid, liquid or gas.
11. T or F—The fire loss control program is divided into five elements.
12. T or F—The atmosphere contains approximately 21% oxygen by volume
13. T or F—The fire tetrahedron does not represent the chain reaction among chemicals.
14. T or F—One of the most important parts of a local loss prevention program is pre-planning for emergencies.
15. T or F—Type B fires deal with electrical fires.

REFERENCES

American Society for Testing and Materials, (ASTM) Standards (published annually).
Bird, Frank E., and Germain, George L. *Practical loss control leadership*, Loganville, GA. International Loss Control Institute, 1992.

Brauer, Roger L. *Safety and health for engineers*, 2nd ed., John Wiley & Sons, Hoboken, NJ, 2006.
Building Officials and Code Administrators International, Inc. BOCA. National Building Codes, 2006.
Fire Protection Handbook 22nd ed., National Fire Protection Association, 2008.
Della-Giustina, Daniel E. *The fire safety management handbook* 2nd ed., American Society of Safety Engineers, 2003.
Della-Giustina, Daniel E. *Safety and environmental management*, Van Nostrand Reinhold, 2007.
National Fire Protection Association, National Fire Codes. Life Safety Code 101, 2008.
Occupational Safety and Health Standards, General Industry (29 CFR-Part-1910).

7 Emergency Management and Preparedness

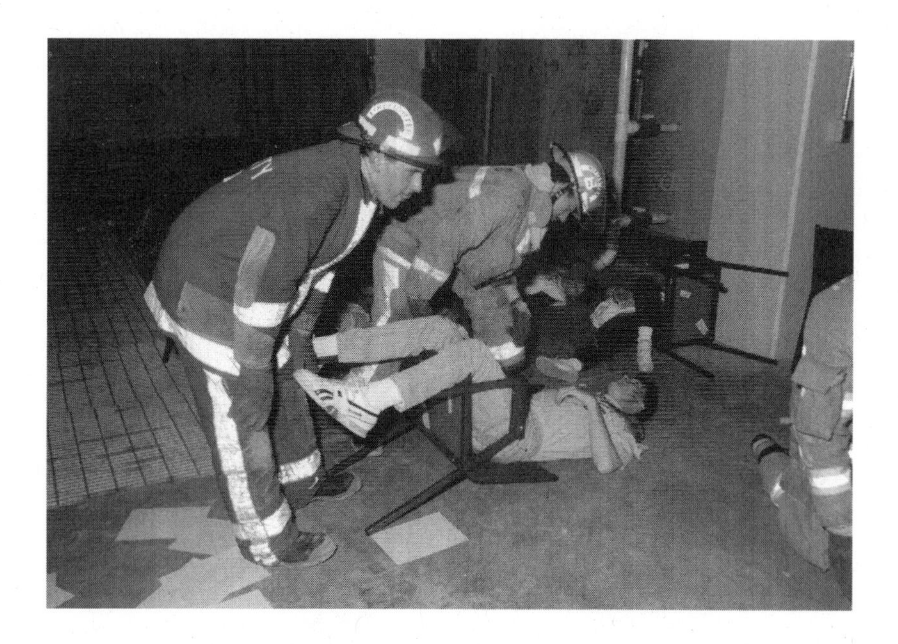

INTRODUCTION

The key to minimizing or controlling the cost and death toll of emergencies is prevention. Crucial to the success of any emergency preparedness plan is preplanning and preparation to develop an effective action plan.

The safety and protection of employees, company assets, and the community in which the company resides is incumbent upon the corporation. Hopefully, an emergency response plan will need to be used only during testing and exercises. However, should an incident occur, a good plan will save time, resources, money, and more importantly, lives. It will also preserve community confidence in the company, which is a very important aspect of successful business.

Today, it is essential to design and develop plans of action in the event of an emergency to ensure the safety and well-being of individuals and property. A disaster is a sudden calamitous event that brings widespread damage or suffering, loss or destruction, and great misfortune, and often arrives without any warning.

Definitions

MSDS: Material safety data sheets
SERP: Site Emergency Response Plan
OES: Office of Emergency Services
EMA: Emergency Management Agency
HAZWOPER: Hazardous Waste and Emergency Response

RESPONSIBILITIES/PROCEDURES

National Incident Management System (NIMS)

The basis for establishing NIMS was to standardize the incident management processes, protocol, and procedures that all emergency responders—federal, state, local, and tribal will implement to coordinate and conduct response actions. The Incident Command System (ICS) was established by NIMS as a standard incident management organization with five functional areas. They are as follows:

- Command
- Operations
- Planning
- Logistics
- Finance/ Administration

This unified command has been universally incorporated into NIMS and provides for and ensures joint decisions, objectives, strategies, plans, priorities, and public communications.

The National Response Plan (NRP) located within the Department of Homeland Security came along with the NIMS. The NRP establishes a comprehensive all-hazards approach to enhance the ability of the United States to manage domestic incidents. It also forms the foundation of how the federal government coordinates with state, local, and tribal governments, as well as the private sector during incidents.

The NIMS Integration Center was established to ensure that NIMS remains an effective and accurate management tool. Again, this falls under Homeland Security to assess proposed changes to NIMS and to capture and evaluate lessons learned as the best practices.

Implementation of this plan should be through the corporation management team as directed by senior management, who designate a coordinator whose primary duties associated with this responsibility include:

- Assignment of emergency responsibilities to employees relative to their competencies and normal work functions.
- Reviewing the plan with the county OES.
- Providing training and preparations for team members and employees.
- Giving commands that carry authority.
- Obtaining and maintaining equipment and supplies necessary for emergency situations.

- Providing for communications and transportation systems during emergencies.
- Advising and leading during simulated drills of emergency plan.
- Periodically evaluating and revising the emergency plan.

DEVELOPING AN EMERGENCY RESPONSE PLAN

Many possible incidents may affect the enterprise, both from within the boundaries of the organization and from without. Some of these incidents may be natural, such as severe thunderstorms, tornadoes, floods, and severe winter weather. Many others may be manmade such as flammables, toxins, reactive gases, fire, power failure, explosion, bomb threat, and hazardous materials accidents.

To be properly prepared to handle such conceivable incidents and to keep them in the realm of emergency rather than disaster, a Site Emergency Response Plan (SERP) with the following nine elements is recommended:

1. A list of emergency numbers for company team members, fire departments, medical and rescue services, and police.

   The first part of the SERP should be a list of emergency phone numbers for fire departments, medical and rescue services, police, and company team members. The list should be short in order to speed response and reduce confusion, but complete enough to ensure proper response. For relevant company personnel, home, work, or cellular telephone numbers should be listed. The situation will dictate which personnel will need to be called and which organizations will need to be contacted. Additional resources may be readily available. They include: local and county emergency management officials, healthcare officials, media contacts, and national groups that provide assistance and response to emergencies and disasters.
2. Site evacuation routes and procedures, both primary and secondary.

   Evacuation procedures should identify both primary and secondary evacuation routes for all hazardous locations, personnel accounting procedures, assembly areas, internal member responsibilities, requirements recommended by local policy, MSDSs, code directives, and manufacturer standards. In addition to the area overview, a specific plan should be outlined for each hazardous location. This will ensure that response is not hindered by irrelevant information. Not only should there be emergency evacuation routes for the facility, but evacuation routes from the facility to other locations in the community are a must, should the need arise to vacate the area in an orderly and efficient manner.
3. Location, type and availability of both site and community emergency equipment.

   This identifies the location, type, and availability of equipment available on site, or from local resources elsewhere. It covers whatever may be needed to bring a situation under control and gives a ready answer for each situation and a path to be followed. Such equipment includes (but is not limited to): firefighting equipment, emergency medical gear, wind speed and direction indicators, communication equipment, self-contained breathing apparatus, and protective apparel.

4. A plot plan designated hazardous materials locations and operations.

   This should be a scaled plan identifying: affected areas within the incident cordon, areas affected due to environmental factors, evacuation routes, and response routes (primary and secondary). It guides the overall response and can be referred to during investigation and recovery procedures as well.

   Scale diagrams and maps need to include the following: existing structures, property lines, entrances and exits, fire hydrants, sprinkler systems, emergency equipment, plant-controlled shut-offs, and hazardous material locations. These locations should be color-coded to ensure ready recognition. Scaled overlays should be drawn to show safe distances. The overlays can be placed over the point of incident to show the cordon area. However, because an incident can extend beyond site boundaries, the plot maps should cover all potential affected areas. This process helps evaluate risk to both the site and the community.

5. Material safety data sheets on all hazardous materials at or near the location.

   MSDSs must be available for quick reference and training. Plant employees, in-house and local response agencies, must understand the hazards of materials used on site. In addition to data on chemical reactions, MSDSs provide firefighting information and telephone numbers for agencies that can provide assistance.

6. A crisis communication plan for dealing with the media.

   Corporate emergencies have a tremendous impact on the communities they are in and command much media interest. A well developed crisis communication plan and well trained personnel are essential to protect the company's reputation and standing in the community. Companies can develop their own plans or get help in developing their crisis communication plans from various consulting firms. A firm that can bring together experts in public relations and the technical field is most advantageous.

   Planning is divided into three categories: content, administration, and logistics. In developing a plan, start with intelligence gathering and include everyone in your organization because they will have a role in the implementation. Remember, during the early stages of a crisis is when you are most vulnerable because you have the least information and it's the time when questions from the media and public begin. Emergency plans and procedures are living documents that continually change and must be enhanced, updated, and revised regularly.

7. Plan coordination, recommendations, and contacts of site and community officials and emergency responders.

   It is important to review coordination plans within the organization and with community and emergency officials, as well as with other neighboring industries (it is suggested that the corporation establish a mutual aid agreement with similar enterprises) to ensure that proper coordination is maintained. The coordination network ensures that all involved have reviewed the plan, provided input, understand specific functions, and agree to those responsibilities. The coordination process should be reviewed periodically and updated as necessary.

8. Training information, including responsibilities of site personnel.

   Training involves organizational members as well as emergency responders outside the company. Employees must be trained in certain tasks, such as power disconnect and use of fire extinguishers, as well as in search-and-rescue or emergency response procedures. MSDSs and chemical properties, emergency reporting procedures, HAZWOPER and specialized equipment, should be covered as well. Supervisors must be trained as team coordinators, onsite commanders, media representatives, etc.. Managers at all levels must be trained regarding program elements so they can effectively serve as liaisons to corporate, regulatory, and local agencies. Such personnel must understand what employee training addresses and know that the SERP defines and assigns responsibilities.

   A place reserved in the SERP is intended to readily give information on specific capabilities of individual team members regarding emergency response activities.

9. Testing dates and procedures, including site exercise results and recommendations.

   This includes drills, tests of various program elements and response capabilities, evaluating response procedures, and corrective actions. Alarm tests, simulated drills, and mock exercises with community groups are several testing approaches. Evaluation results and proposed or actual corrective actions must be documented and incorporated into the plan. Involving community agencies in the testing process enhances community relations and improves the plan. It can also lead to improved training opportunities.

Reports, Forms, and Record keeping

Proper record keeping, maintaining accurate reports, and forms used, help in reviewing past performances in drills, exercises, and events. Records of what was done and when, such as training, exercises, etc. not only serve as a record of what you have done for the sake of preparedness, but help keep you current by reminding you that time for periodic exercises or refresher training is due.

Training for Success

Training is an essential part of most activities within any organization. This is especially true when it comes to emergency response and action. People should be trained to handle emergency situations within their workplace. Proper training in emergency response will prove a valuable asset to the company as well as to the employees should the need arise.

Such training may include company team members as well as emergency responders outside the company. Training for employees should include: power disconnects, the use of fire extinguishers, search-and-rescue techniques, emergency response procedures, emergency first aid or medical treatment, and more.

Training in emergency medical treatment may be available through local hospitals or fire departments. Certification of training is usually available through these organizations as well. Benefits will include a closer working relationship with local fire departments and emergency response community. Training for company employees (such as fire brigades) may be acquired free of charge from local fire departments. Employees may attend certain training sessions held at fire department facilities, or fire department personnel may come to your enterprise and provide essential training for all employees.

Training should include MSDS and chemical properties, emergency reporting procedures, HAZWOPER and specialized equipment. Supervisors must be trained as team coordinators, onsite commanders, media representatives etc. Managers at all levels must be trained regarding program elements so they can effectively serve as liaisons to corporate, regulatory, and local agencies. Such personnel must understand what employee training addresses and know that the SERP defines and assigns responsibilities. The top executive at the site serves as an approving agency of the plan and must be included in this process.

Hazardous Materials

On many sites there is the possibility of several thousand chemicals and other hazardous materials that could affect not only the employees but also the surrounding community. These effects could range from respiratory troubles, skin irritation, eye injury and irritation, contamination of the community, explosions, and fires. This makes it every important to know what is on your site and also what materials are present on neighboring sites.

The location of hazardous materials needs to be known so that emergency personnel will know where the materials are located and the type of substance they are dealing with. It would be a good idea to have a diagram of the facility with the locations and names of hazardous substances plotted on the diagram.

Public Demonstrations/Civil Disturbances

In recent years we have seen a variety of demonstrations for a multitude of different reasons. Some demonstrations develop slowly, allowing for negotiations, crowd control measures, and time for authorities to assess the problem. Yet, on other occasions, problems can spring up with little or no advance warning.

In situations where the situation is not rapidly developing you can go ahead and operate out of your normal offices and hope the negotiating works out the problem and everything can return to normal in a short time.

In a situation in which a sudden outbreak, often involving violence, takes place, it is important to work with law enforcement agencies to resolve the problem and to protect the employees and property of the company. The law enforcement agency can help in making decisions as to whether to continue to operate the facility.

Sabotage

No company or facility is immune to sabotage. However, the types and targets for sabotage can be predicted. The person or group behind the sabotage will look for a target that is easy to access, is critical to the operation of the facility, and at least partially self destructible. Those carrying out the sabotage are possibly enemy agents, disgruntled employees, or individuals who are mentally ill.

Terrorism

Terrorism is a covert and criminal act that provides problems for management and emergency service personnel. Many of these acts of terrorism deal with bomb incidents, bomb threats, and the taking of hostages. To be prepared, contact must be made with the local enforcement agencies, the FBI, and bomb disposal units. This allows for the assistance of more experienced personnel.

Experience as shown that 95% of all bomb threats are hoaxes. However, such a threat may be authentic. Appropriate action should be taken to provide for the safety of employees and the community even if it turns out to be a hoax.

If a suspicious object is spotted and thought to be a bomb the local authorities need to be notified and a bomb disposal unit contacted to remove the object.

REVIEW QUESTIONS

True or False

1. T or F—There is no one basic emergency and disaster plan that fits all facilities and operations.
2. T or F—Material Safety Data Sheets (MSDS) have to be located in the main office.
3. T or F—A plot plan consists of hazardous material locations and operations.
4. T or F—During an emergency a complex legalistic disaster plan would be your best resource.
5. T or F—The responsibility for coping with natural disasters rests with the Red Cross.
6. T or F—Mine disasters are normally a result of a buildup of methane gas or coal dust that ignites.
7. T or F—Proper record keeping, maintaining accurate reports and forms used helps in reviewing past performances in drills, exercises, and events.
8. T or F—Two elements are recommended as part of a Site Emergency Response Plan (SERP).
9. T or F—Planning is divided into three categories: content, administration, and logistics.
10. The first part of the ______ ______ ______ ______ should be a list of emergency phone numbers for fire departments, medical, police, and rescue services.
11. ______________ pressure is associated with tornado damage.

12. Experience has shown that _____________% of all bomb threats are hoaxes.

REFERENCES

Aysan, Y and I. Davis, Eds. *Disaster and the small dwelling*. London: James Science, 1992.

Della-Giustina, Daniel E. *Safety and environmental management.* Lanham, MD: Roman and Littlefield. 2007.

Della-Giustina, Daniel E. *Motor fleet safety and security management.* Boca Raton, FL: CRC, 2004.

Disaster Preparedness Manual, West Virginia Safety and Environmental Management Department, 2007.

Hospitals and Community Emergency Response—What you need to Know, OSHA 3152, 1997.

Scholl, Craig R. *Industrial fire protection handbook*, Technomic, 1992.

Molino, Louis N. Sr. *Emergency incident management systems fundamentals and applications.* New York: John Wiley, 2006.

Della-Giustina, Daniel E. *Planning for school emergencies*, Waldorf, MD: AAHPERD Publications, 1988.

Vulpitta, Richard J. *On Site emergency response planning guide.* Itasca, IL: National Safety Council, 2003.

8 Logout/Tagout

INTRODUCTION

The purpose of these instructions is to ensure that, before any employee performs servicing or maintenance on machinery or equipment, where the unexpected energizing, start-up, or release of any type of energy could occur and cause injury, the machinery or equipment will be rendered safe to work on by being locked out or tagged out.

When an employee performs work on machines, equipment, or systems, and is exposed or causes other employees to be exposed to the hazards of unexpected energization, these procedures must be followed.

Definitions

Affected Employee: An employee whose job requires him or her to use or operate a machine or equipment on which servicing or maintenance is being performed under lockout or tagout, or whose job requires working in an area where such servicing or maintenance is being performed. All employees in the facility are considered to be affected employees.

Authorized Employee: An employee who locks or tags out machines or equipment in order to perform the servicing or maintenance on that machine or equipment.

Clear: To prepare equipment by isolating and placing it in a safe condition to avoid injury or damage if equipment should unexpectedly start during the "Try" step. This includes removing and warning personnel.

Energized: Connected to an energy source or containing residual or stored energy.

Energy Isolating Device: A mechanical device that physically prevents the transmission or release of energy. The term does not include a push button, selector switch, and other control circuit type devices.

Energy Source: Any source of electrical, mechanical, hydraulic, pneumatic, chemical, thermal, or other energy.

Hot tap: A procedure used in the repair, maintenance, and service activities that involve welding on a piece of equipment under pressure in order to install connections or appurtenances. If hot tap operations are performed at a location, a supervisor for the hot tap must be appointed and an instruction sheet for the hot tap prepared.

Instructions: The document specifically prepared for specific types of equipment containing such information as the energy-isolating devices and their identification, type of energy-isolating devices, presence and location of any stored energy, any appropriate procedure for lockout, supervisor's name, authorized employees' job titles, and equipment location. An instruction sheet will be prepared for each type of machinery or equipment at the location covered under this policy. Instruction sheets do not need to be prepared in certain limited circumstances, which are listed in the procedures section of this document. Preparation of the instruction will be the responsibility of the supervisor, who has been assigned responsibility for the machinery or equipment.

Lockout: The placement of a lockout device on an energy-isolating device, in accordance with an established procedure, ensuring that the energy-isolating device and the equipment being controlled cannot be operated without removal of the lockout device.

Lockout device: A device that utilizes a positive means such as a lock to hold an energy-isolating device in the safe position and prevent the energizing of a machine or equipment. This device shall be:

- Used only for controlling energy, not other purposes
- Durable
- Standardized by color, shape, or size
- Substantial enough to prevent removal absent excessive force
- Able to indicate the identity of the employee applying the device(s)

Servicing and or Maintenance: Workplace activities such as constructing, installing, setting up, adjusting, inspecting, modifying, and maintaining, or servicing machines or equipment. This includes lubrication, cleaning or unjamming of machines or equipment and making adjustments or tool changes where the employee may be exposed to the unexpected energization or start-up, or release of hazardous energy.

Supervisor: The supervisory or management person appointed with responsibility for the specified machine(s) or equipment.

Try: To test equipment to ensure the effectiveness of the lockout and the removal of stored energy.

Responsibilities/Procedures

Responsibility

Supervisors and authorized employees shall receive instruction on lockout procedures. Affected employees shall receive instruction on the purpose and use of the lockout procedure. Area supervisors and the authorized and affected employees are to be listed on the specific instruction sheets for each type of machine or equipment.

Preparation

Authorized employees should obtain the appropriate instruction sheet for the machine or equipment on which work will be performed. These instruction sheets shall be located in the area office for the area where the machine or equipment is located, with duplicates maintained by the safety manager. He or she should make a survey to locate and identify all isolating devices to be certain which switch, valve, or other isolating devices apply to the equipment to be locked out and where, if any, stored energy is located. (Multiple energy sources may be involved.)

Sequence of Lockout Procedure

The authorized employees will:

Notify all affected employees in the area that a lockout device is going to be utilized and the reason for that prior to installation of the lockout device.

Know the type and magnitude of energy that the machine or equipment utilizes and shall understand the hazards they could present.

If the machine or equipment is operating, shut it down using the normal stopping procedure.

Position the switch, valve or other energy-isolating device so that the equipment or machine is isolated from its energy source. Stored energy (such as that found in springs; elevated machine members; rotating flywheels; hydraulic systems; and air, gas, steam, or water pressures, etc.) must be dissipated or restrained by methods such as repositioning, blocking, bleeding, etc.

Lockout the energy isolating devices with assigned individual locks and identification tags.

After ensuring that no personnel are exposed, and as a check on having disconnected the energy sources, operate the push button or other normal operating controls to make certain the equipment will not operate.

When all steps are satisfactorily completed, return all operating controls to the "neutral" or "off"" position after the test to assure no inadvertent starting or releasing of energy later during deactivation of the lockout.

The machine or equipment is now locked out.

Lockout Deactivation

When the servicing or maintenance activity is complete and the equipment or machine is ready for normal operations, the authorized employee will:

Check the area around the equipment or machines to ensure that no one is exposed.

Notify all affected employees in the area before the lockout device is removed.

After all tools have been removed from the machine or equipment, guards have been reinstalled, and all employees are in the clear, remove the lockout devices.

Return the energy-isolating device to its normal position, restoring energy to the machine or equipment.

Temporary Removal of Lockout Device during Servicing

In the event a lockout device must be temporarily removed and the machine or equipment energized to test or position the machine, the authorized employee shall ensure safety in the following manner:

Clear the machine or equipment of tools or materials and remove employees from the area.

Remove the lockout device.

Energize the machine or equipment and proceed with testing or positioning.

De-energize all systems and reapply energy control measures by following the lockout procedure to continue work.

Compliance of Outside Personnel

When outside servicing personnel are involved in activities covered by this procedure, the outside employer shall be informed about this procedure and shall inform the safety manager of its lockout or tagout procedure. If such personnel fail to follow the procedure, they must be instructed to follow it. Subsequent failure to follow this procedure should be enforced by stern measures, up to and including removal of the personnel from the facility.

Removal of a Lockout Device in the Authorized Employee's Absence

If the authorized employee is not available to remove the lockout device, it may be removed only under the following circumstances:

Upon certifying that the authorized employee is not at the facility.
After reasonable efforts have been made to contact the authorized employee to inform him or her of the need to remove the device(s).
After ensuring the authorized employee has knowledge of the removal before work is resumed.

Exceptions to Lockout Procedure

The lockout procedure must be followed for work on any machinery, except when the following conditions occur:

Normal production operations, as long as guards or other protective devices are not deactivated or bypassed. If a guard or protective device is removed or deactivated, the machinery must be locked out before any work is performed. If machinery must be energized at some point during the service procedure, the lockout device may be removed only for the duration necessary to perform that service procedure. In these instances, employees must use special control circuits, equipment, or operating procedures available that are designed to provide effective protection from the machinery danger zone. For any machine where it is necessary to perform a service procedure with the machine energized, the special control circuits, equipment, or procedures to be utilized will be listed on the inspection sheet for that machine.
Work on cord- and plug-controlled equipment where energization is controlled by unplugging the equipment and the plug remains under the exclusive control of the employee performing the work.
Hot tap operations involving gas, steam, or water lines where continuity of operation is essential and shutdown is impractical. In this case, established procedures for the specific hot tap operation (as listed on the applicable instruction sheet) must be followed and special equipment designed to protect employees from the zone of danger must be used.

Exceptions to Instruction Sheet Requirements

An instruction sheet is not necessary if all of the following conditions are met:

The machine or equipment has no potential for stored or residual energy or reaccumulation of stored energy after shutdown, which could endanger employees.
The machine or equipment has a single energy source that can be readily identified and isolated.
The isolation and lockout of the energy source will completely de-energize and deactivate the machine or equipment.
The machine or equipment is isolated from that energy source and locked out during servicing or maintenance.
A single lockout device will achieve a "locked-out" condition. The servicing or maintenance creates no hazard for other employee.

However, if an accident occurs that involves the unexpected activation or reenergization of the machine or equipment during servicing or maintenance where no instruction sheet was provided, the location must immediately implement a written instruction sheet for the machinery or equipment.

Types of Lockout

Individual Lockout

Each employee working on equipment will follow the Lock, Clear, and Try steps according to instructions. The employee will retain the key to the lock in his/her possession. Only the employee locking the machine is authorized to remove the lock, except under the special circumstance noted earlier.

Group Lockout

If more than one individual is required to lockout equipment, one of the two group lockout procedures shall be followed:

1. *Group Lockout of the Isolating Device*—The authorized employee in charge of the group is responsible for ensuring the Lock, Clear, and Try steps are followed. The authorized employees in the group are responsible for verifying these steps. Each person performing work on the equipment shall place their own personal lockout device on the energy isolating device. When an energy isolating device cannot accept multiple locks, a multiple lockout device (hasp) may be used. As each person completes work and no longer needs lockout protection, they shall remove their lockout device.
2. *Group Lock Box*—The authorized employee in charge of the group is responsible for ensuring the Lock, Clear, and Try steps are followed. The authorized employees in the group are responsible for verifying these steps. The only lockout device(s) affixed to the energy-isolating device will be that of the authorized employee in charge. The key(s) to these lockout device(s) shall be placed in a group lock box. The authorized employee in charge of the group shall initially place his or her personal lockout device onto the box. As each person completes work and no longer needs lockout protection, he or she will remove his or her lockout device.

Shift Change Lockout

During shift changes, the continuity of lockout protection must be ensured.

Any time a machine needs to be locked out across a shift change, the oncoming authorized supervisor shall ensure that his or her lockout device is removed. In the event this sequence is not followed and the lockout device is removed before the oncoming employee arrives, the oncoming employee must follow the entire lockout procedure before beginning work.

Fixed Electrical Equipment and Circuits: Working on or Near Exposed De-energized Parts

Application

This applies to work on exposed de-energized parts, or near enough to them to expose the employee to any electrical hazard they present. Conductors and parts of electrical equipment that have been de-energized but have not been locked out or tagged in accordance with this paragraph shall be treated as energized parts with 29 CFR 1910.333(c) (OSHA, General Industry) applying to work on or near them.

Lockout and Tagging

When any employee is exposed to contact with parts of fixed electrical equipment or circuits that have been de-energized, the circuits energizing the parts shall be locked out or tagged out or both in accordance with the requirements of this paragraph. The requirements shall be followed in the order in which they are presented.

NOTE: Fixed equipment in this section refers to equipment that is fastened in place or connected by permanent wiring methods.

Procedures

A written copy of this procedure is available for inspection by employees and by the assistant secretary of labor and his or her authorized representatives.

De-energizing Equipment

Safe procedures for de-energizing circuits and equipment shall be determined before circuits or equipment are reenergized.

The circuits and equipment to be worked on shall be disconnected from all electric energy sources. Control circuit devices, such as push buttons, selector switches, and interlocks, may not be used as the sole means for de-energizing circuits or equipment. Interlocks for electric equipment may not be used as a substitute for lockout and tagging procedures.

Stored electric energy that might endanger personnel shall be released. Capacitors shall be discharged and high capacitance elements shall be short-circuited and grounded, if the stored electric energy might endanger personnel.

Stored non-electrical energy in devices that could reenergize electric circuit parts shall be blocked or relieved to the extent that the circuit parts could not be accidentally energized by the device.

Application of Locks and Tags

A lock and tag shall be placed on each disconnecting means used to de-energize circuits and equipment on which work is to be performed. The lock shall be attached so as to prevent persons from operating the disconnecting means unless they resort to undue force or the use of tools.

Each tag shall contain a statement prohibiting unauthorized operation of the disconnecting means and removal of the tag.

If a lock cannot be applied, a tag may be used without a lock.

A tag used without a lock shall be supplemented by at least one additional safety measure that provides a level of safety equivalent to that obtained the use of a lock (i.e. removal of an isolating circuit element, blocking off of a controlling switch, or opening of an extra disconnecting device).

A lock may be placed without a tag under the following conditions:

- Only one circuit or piece of equipment is de-energized.
- The lockout period does not extend beyond the work shift.
- Employees exposed to the hazards associated with reenergizing the circuit or equipment are familiar with this procedure.
- The lock should be used with a tag under normal circumstances.

Verification of De-energize Condition

The requirements of the following shall be met before any circuits or equipment can be considered and worked on as deenergized.

A qualified person shall operate the equipment operating controls or otherwise verify that the equipment cannot be restarted.

A qualified person shall use test equipment to test circuit elements and electrical parts of equipment to which employees will be exposed and shall verify that the circuit elements and equipment parts are deenergized. The test shall also determine if any energized conditions exists as a result if inadvertently induced voltage or unrelated voltage feedback occurs even though specific parts of the circuit have been deenergized and presumed to be safe. If the circuit to be tested is over 600 volts, nominal, the test equipment shall be checked for proper operation immediately before and immediately after this test.

Reenergizing Equipment

These requirements shall be met, in the order given, before circuits or equipment are energized, even temporarily.

A qualified person shall conduct tests and visual inspections as necessary to verify that all tools, electrical jumpers, shorts, grounds, and other similar devices have been removed so that the circuits and equipment can be safely energized.

Employees exposed to the hazards associated with reenergizing the circuit or equipment shall be warned to stay clear of circuits and equipment.

Each lock and tag shall be removed by the employee who applied it or under his or her direct supervision. However, if this employee is absent from the workplace, then the lock or tag may be removed by a qualified person designated to perform this task provided that:

- The employer ensures that the employee who applied the lock or tag is not available at the workplace.
- The employer ensures that the employee is aware that the lock or tag has been removed before he or she resumes work at that workplace.
- There shall be a visual determination that all employees are clear of the circuits and equipment.

Training and Certification

Initial training shall be provided to ensure that the purpose and function of the energy control program is understood by all employees.

The knowledge and skills required for the safe application, usage, and removal of energy controls is known.

All authorized employees shall receive training in the recognition of applicable hazardous energy sources, the type and magnitude of the energy available in the workplace, and the methods and means for isolation control. Thereafter, any employee who becomes an authorized employee shall receive the appropriate training at the time of becoming authorized.

New employees will receive the appropriate training as part of the orientation process. This item will be added to the standard orientation checklist.

Retraining

Retraining shall occur for all authorized and unauthorized employees whenever there is a change in their job assignments; a change in machine, equipment, or processes that present a new hazard; or when there is a change in the energy control procedure.

Additional retraining shall occur whenever a periodic inspection reveals, or whenever the employer has reason to believe, that there are deviations from or inadequacies in the employee's knowledge or use of procedures.

The safety manager is responsible for certifying that training has occurred and is being kept current. The certification shall contain the name of the employee and the training date. The safety manager will also keep an outline of the training sessions conducted.

Inspections

Periodic inspections of the energy control procedures shall be done at least annually to ensure procedures and requirements are met and to correct any deviations or inadequacies.

The inspection shall include a review between the inspector and each authorized employee of the employee's responsibilities under the energy control procedure.

The inspection shall be done by an authorized employee other than those utilizing the energy control procedures being inspected. The safety/training manager shall have responsibility for these inspections.

The safety/training manager shall certify that the periodic inspections have been completed. This certification shall identify the machines or equipment on which the energy control procedure was utilized, the date of the inspection, the employees included, and the person performing the inspection.

REVIEW QUESTIONS

1. ______________ is a mechanical device that physically prevents the transmission or release of energy. The term does not include a push button, selector switch and other control circuit type devices.
2. ______________ and ______________ employees shall receive instruction on lockout procedures. ______________ employees shall receive instruction on the purpose and use of the lockout procedure.
3. A ______________ and ______________ shall be placed on each disconnecting means used to deenergize circuits and equipment on which work is to be performed.
4. Retraining shall occur for all ____________ and ____________ employees whenever there is a change in their job assignments; a change in machine, equipment or processes that present a new hazard; or when there is a change in the energy control procedure.
5. The ______________ Manager shall certify that the periodic inspections have been completed.
6. Periodic inspections of the ______ ______ ______ shall be done at least annually.
7. A ______________ ______________ shall conduct tests and visual inspections.
8. 29 CFR 1910.147 is the standard on ______ ______________ of _______ ______________ (Lockout/Tagout).

REFERENCES

Code of Federal Regulations

29 CFR 1910.147—OSHA General Industry Standard: The control of Hazardous Energy
29 CFR 1910.333—OSHA General Industry Standard: Use Code of Federal Regulations, Title 29 CFR 1910.147

BIBLIOGRAPHY

National Safety Council (NSC). Accident prevention manual for business and industry, 10th edition, Itasca IL, 1992.
Occupational Safety and Health Administration, The control of hazardous energy.
(Lock/Tagout) Standard, 1910.147 U.S. Department of Labor, Washington D.C., 1998.
Della-Giustina Daniel E. Safety and environmental management, New York: Van Nostrand Reinhold, 1996.
National Fire Protection Association. The national electrical code handbook, Quincy MA, 2008.

9 Confined Space Entry

INTRODUCTION

Confined spaces on first appearance may seem relatively hazard free, but contain many hazards that can pass as a threat to entry personnel. It is critical that each confined space in your facility be evaluated individually to determine that all hazards are identified. The following categories cover the major hazard groups found in confined spaces:

- Atmospheric
- Content issues
- Potential energy
- Environment in the space
- Configuration of the space
- Nature of the work
- External hazards
- Miscellaneous

In addition to these hazards, which are unique to confined spaces, one must also contend with the general safety hazards that are present in any work environment. The evaluation must include an overall assessment of all of the potential hazards within the space. Ideally, hazards should be eliminated. If this is not possible, one must control the hazards to a level of acceptable risk. The corporation's employees must be aware of the general characteristics of confined spaces and how to recognize a work area that may not have been posted as a confined space. When doing repair maintenance work outside the plant, each service employee must be capable of recognizing and identifying a confined space, and take the proper procedures in dealing with the danger.

The specific methods described in this sample written program are for illustrative purposes, and other effective methods may be substituted to satisfy local needs or practices.

Definitions

Confined Space: Any enclosed or partially enclosed equipment or area that:
Is not designed for normal occupancy by employees.
Has limited or restricted access.
Is large enough to provide bodily entry.

Permit Required Confined Space: A confined space that has one or more of the following characteristics:
Contains or has the potential to contain hazardous material.
Contains a material that could bury or entrap employees.
Has an internal design with inward sloping walls or floors to a smaller area.
Contains a recognized serious safety or health hazard.

Non-Permit Confined Space: A confined space that does not contain an actual or potential hazardous atmosphere. Any hazard that can cause death or physical harm.

Entry: Occurs when employees place any part of their bodies inside the confined space or opening.

Attendant: A trained employee assigned to stand by outside the confined space entry opening to observe the employee inside the space, to assist the employee inside, and provide help and retrieval if needed.

"Blind" or "Blinding" or "Blanking": The absolute closure of a pipe, line, or duct, to prevent passage of material (e.g., by fastening a solid plate or "cap" across the pipe).

Engulfinent: The surrounding and effective capture of a person by a liquid or finely divided (flowable) solid substance that can be aspirated to cause death by filling or plugging the respiratory system or that exert enough force on the body to cause death.

Entrant: Any employee who is authorized to enter a permit-required confined space.

Entry Permit: The employer's written authorization to allow and control entry into a permit space under defined conditions for a stated purpose, and during a specified time.

Entry Supervisor: An individual responsible for determining if acceptable conditions are present at a permit space where entry is planned, for authorizing entry, overseeing entry operations, and for terminating entry.

Hazardous Atmosphere: An atmosphere that may expose an employee to risk of death, disablement, impairment of ability to self-rescue, injury, or acute illness from one or more of the following causes:

Flammable gas, vapor, or mist in excess of 10% of its lower explosive level (LEL).

An oxygen-deficient atmosphere containing less than 19.5% oxygen by volume or an oxygen-enriched atmosphere containing more than 23% oxygen by volume.

Airborne combustible dust at a concentration that meets or exceeds its lower flammable limit (LFL).

An atmospheric concentration of any substance above the listed numerical value of the permissible exposure limit (EL). Any other atmospheric condition that is immediately dangerous to life or health.

At concentrations lower than the LEL, the mixture is too lean to burn.

Immediately Dangerous to Life or Death: Any condition that poses an immediate threat to life, or which is likely to result in acute or immediately severe health effects.

Immediate Severe Health Effects: An acute clinical sign of serious exposure-related reactions is manifested within 72 hours of exposure.

Qualified Person: A person who is trained to recognize the hazard(s) of a confined space and how to evaluate those anticipated hazards. The employer may designate an employee as his or her representative for the purpose of assuring safe confined space entry procedures and practices at a specific site.

Retrieval line: A line or rope secured at one end of worker's safety belt, chest, or body harness; or wristlets with the other end secured to an anchor point or lifting device located outside the entry portal.

Zero Mechanical State: The mechanical potential energy of all portions of the machine or equipment at its lowest practical value so that the opening of the pipe(s), tube(s), hose(s), or actuation of any valve, lever, or button will not produce a movement that could cause injury.

RESPONSIBILITIES/PROCEDURES

General Requirements

Confined spaces must be identified. A list should be maintained by the company.

Confined spaces should be identified with signs indicating "Danger, Confined Space Permit Required" or "Confined Space" if a non-permit confined space.

Any space found meeting the definition of a confined space or suspected to be a confined space but not on the list or not identified by a sign, must be considered a permit-required confined space until evaluated.

Permits for confined space entry may be issued only by supervisors or other designated employees trained in these procedures and familiar with the permit process.

Any exceptions to this procedure and the permit must be referred to the environmental health and safety supervisor for review.

No employee may enter a confined space or serve as an attendant until confined space entry training has been completed and documented.

Employees with existing medical conditions likely to be affected by confined space entry should be reviewed by the company's medical doctor, or another qualified person's determination will be made for a physician's evaluation.

Entry into confined spaces should be prohibited in the presence of:

A bin or hopper with material at or above the employee's head.

A confined space with an unsecured hazard.

A confined space with an unknown hazardous atmosphere.

Entry may be made into an atmosphere containing less than 19.5% oxygen for essential emergency repairs only when wearing supplied air system.

The confined space entry procedure should be evaluated annually by the health and safety department and revisions made as needed.

PERMIT

Entry Procedure for Permit-Required Confined Spaces

Supervisor must confirm entry is necessary (work cannot be done from outside).

Supervisor or authorized employee must evaluate the confined space to be entered.

All electrical power to the confined space including that for internal mechanical components must be shut off, locked out, or tagged.

Conveyors, lines, and equipment that feed the confined space must be shut down, blanked or blocked, and locked out or tagged.

Heating systems serving the confined space must be shut down and the gas supply valves locked out and the lines or pipes blanked, blocked, or otherwise secured.

Tanks containing water and sludge must be drained, flushed, and as much sludge as possible removed before entry.

Hatches, doors, and manholes must be opened to provide maximum ventilation.

Water line valves to water system must be closed and secured.

The confined space must be visually inspected from outside before entry to identify hazardous conditions and accumulated material.

Using a calibrated gas monitor, the oxygen and flammable gas levels must be monitored from outside the space and it must be verified that the instrument is not damaged and calibration date is current.

Readings must be taken at 5-foot intervals from top (vertical) entry or side (horizontal) entry if possible.

If oxygen content is below 19.5%, entry is prohibited except for essential emergency repairs using an air supply unit; entry is prohibited if oxygen content exceeds 23.5%.

If the monitor for flammable gas indicates 10% of the LEL or greater, entry is prohibited.

If ventilation is used to increase oxygen or remove harmful gases, the space must be retested using the same procedure as that prior to entry.

Portable monitors must be taken into confined spaces and the attendant must monitor conditions remotely.

The work to be performed must be reviewed including:

Hot work (e.g., welding) and if so, a hot work permit issued.

Is there electrical equipment in or near damp areas, and if so, is it connected to a ground fault circuit interrupter?

Determine if work may affect the atmosphere in the confined space and determine how to control at safe levels.

Personal protective equipment for the confined space and work to be done must be determined and issued.

The means for entry and exit including retrieval must be determined, and a safety harness with D-ring and safety line must be worn for all entries unless the line will create a hazard.

Unless rigging cannot be accomplished or is not needed for retrieval, the safety line must be attached to a manual hoist.

Employees in the personnel hoist must wear safety harnesses attached to a safety line or cable.

All entry equipment must be inspected prior to use.

A trained attendant must be assigned to the entry opening when an employee is in the confined space, and the attendant may not leave the area unless replaced by a trained attendant.

The attendant will have no other duties other than as an attendant.

When the confined space has been evaluated and all conditions for entry have been met, the attendant and employees entering the space will sign their names on the permit. The supervisor will also sign the permit, authorizing entry date and time, and leave the permit at the entry site.

After completion of work in the confined space, the supervisor will:

Confirm that materials, equipment, and all employees have been removed.

Have the entry point and all other openings closed.

Authorize removal of locks, tags, and line blanks and opening of valves.

Sign and date the work completion line on the permit, and return the permit to filing.

Non-Permit Confined Space Entry

The following steps must be taken prior to entry work in a non-permit confined space.

The supervisor will evaluate the area to determine that there are no adjacent operations affecting the confined space.

Test the atmosphere in the space. Verify a safe means for entry.

Verify that work to be done in the confined space will not create hazards for those in the confined space.
Notify employees in the adjacent area of the entry and work to be done. Assign only employees who have had confined space entry orientation or confined space entry training.
Provide employees with specific safety and work instructions.
Suspend work if conditions change which would make the space permit required confined space.

Reclassification of Permit-Required Spaces

A permit-required space may be reclassified to a non-permit confined space by a trained supervisor after:

Confirmation through multiple, separate air sampling that there is no actual or potential hazardous air.
Elimination or securing all hazards in a space.
Completion of documentation showing date, specific space and location, and signature of supervisor making the determination.

If any hazard occurs in the space, the space must be reevaluated to determine if the classification should be changed to permit-required.

Retrieval and Rescue

The attendant is responsible for initiating the retrieval or emergency rescue plan when:

There is reason to believe employees are in danger from inside or outside conditions.
Oxygen/gas readings reach unacceptable levels.
Entry employees are incapable of exiting the space independently.
The attendant is requested by employee to provide aid in the confined space.
The attendant cannot make or confirm voice contact with entry employees.

If assistance will be needed to remove the employee, it should be immediately available (on same floor or general area) and a communication system established. If the attendant determines there is an injury, medical condition, or other health exposure, medical assistance will be requested by radio. If retrieval cannot be made by harness or safety line, the attendant will call for assistance and wait for it to arrive. Attendants and other company personnel will not enter confined spaces for rescue.

Attendant and others who may assist will be trained in and know:
Safety line, harness, and winch are primary means for removal/rescue.
Establish removal/rescue system.
Use of CPR and AED (automated electronic defibrillator).

TRAINING

All employees should receive a confined space orientation, which will include:
What are confined spaces?
How they are marked/identified in the plant?
Entering permit-required confined spaces is prohibited unless the employee is trained and a permit is issued.
Employees may be assigned as an attendant only if trained.
Staying out of entry area during confined space work.
Symptoms of over exposure to gas and lack of oxygen.
Entry into a confined space to rescue people is prohibited.
Prior to initial assignment to confined space work, each employee must be trained in the following:
General hazards of confined spaces.
Entry procedures and permit system including energy isolation/lockout.
Types of confined spaces and their specific hazards.
Communication methods.
Duties of attendants and those entering confined spaces.
Required PPE.
Retrieval/rescue system, including use of hoist, safety harnesses, and safety lines.
Special hazards/work procedures for tools and work to be used in confined space.
Use of portable oxygen and gas monitoring equipment.
Supervisor will be trained in all requirements for confined space entry and know:
All details for properly completing/issuing confined space entry permits.
Duties before, during, and following confined space work.
Specific hazards and precautions for confined spaces under supervisor's direction.
Use of oxygen and gas monitoring equipment.

FINAL REPORT AND DOCUMENT

After an entry has occurred and the person who entered the confined space has completed his or her work, a final report should be written. This report must include:

A copy of the signed permit.
Reason(s) that this confined space had to be entered.
A list of alternative methods explored and reasons that these alternative methods were determined "unfavorable."
Any issue(s) that may come up during the preparation stage.
Any issue(s) that may have come up with the condition of the safety equipment involved.

A copy of this report should be turned into the safety department along with the permit.

Non-Routine Tasks

When performing non-routine tasks, the following need to be looked at before entering the confined space:

- A review of any MSDS for chemical materials being taken into the confined space area
- The need for additional atmospheric monitoring
- Ventilation and other controls

The facility safety manager, other internal personnel, or other resources may be contacted for additional input. If the hazards associated with the non-routine cannot be adequately controlled, entry should not be permitted.

Equipment

- Equipment dedicated only for confined space entry, including oxygen/gas monitors, safety harnesses, personnel hoist, and safety lines will be properly stored and secured when not in use.
- Prior to each entry, all equipment will be visually inspected for damage or other defects. Damaged or defective equipment will not be used and will be bagged for repair or discarded.
- Following use, entry equipment will be cleaned and stored.
  - Oxygen/gas monitors must have their calibrations checked per manufacturer's recommended intervals.
  - Calibrations must follow the manufacturer's instructions with calibration unit supplied.
  - Calibrations should be made by trained personnel in engineering and maintenance.
  - Safety harnesses for confined space entry must have D-ring and leg straps. Waist belts are not acceptable for confined space entry.
  - Safety lines must have a minimum breaking strength of 5,400 lbs and can be made of manila, nylon, or specially designed line for rescue but must be compatible with chemicals to which they may be exposed. They must be inspected before use.
- The hoist for lowering and raising an employee can be either manually operated or electrically powered and:
  - The hoist must be stable and all components capable of handling five times the weight of the person to be lowered/raised (minimum 1,500 lbs).
  - The hoist must have a brake to prevent reverse travel when raising and individual.
  - Cables, hooks, and other components must be inspected per manufacturer's recommendations.
  - Lighting and electrical equipment must be acceptable for the area and conditions were used. Lights and power cords must be acceptable for use in wet damp locations.

Electrically powered tools and equipment and lighting used in damp confined spaces must be connected to a ground fault circuit interrupter (GFCI).

Power ventilation providing fresh air to confined spaces must have make-up air to the blower supplied from a point where there are no other known harmful air contaminants and:

Exhaust blowers should not discharge into employee work area.

Exhaust blowers removing flammable gases must not discharge near sources of ignition.

REVIEW QUESTIONS

TRUE OR FALSE

1. T or F—Oxygen/gas monitors must have their calibrations checked per manufacturer's recommended intervals.
2. T or F—Calibrations should be made by trained personnel.
3. T or F—Electrically powered tools and equipment and lighting used in damp confined spaces must not be connected to a ground fault circuit interrupter (GFCI).
4. T or F—When performing non-routine tasks you need to look only at MSDS before entering the confined space.
5. T or F—Any employee is authorized to enter a confined space.
6. T or F—All employees should receive a confined space orientation.
7. T or F—The hoist for lowering and raising an employee can only be manually operated.
8. T or F—Prior to each entry, all equipment will be visually inspected for damage or other defects.
9. T or F—The entry employee is responsible for initiating the retrieval or emergency rescue plan.
10. T or F—All confined spaces start off as non-permit required confined space.

BIBLIOGRAPHY

Roughton, J. Confined space entry: An overview, *Journal of Professional Safety*, 1993.

Pettit, T. H. Linn. A guide to confined spaces, U.S. Department of Health and Human Services, NIOSH, Division of safety Research, Morgantown, 1987.

OSHA. Permit-required confined spaces OSHA 3138, 1998.

Della-Giustina, Daniel E. *Safety and environmental management*, Lanham, MD: Roman and Littlefield, 2007.

10 Personal Protective Equipment

INTRODUCTION

OSHA Personal Protective Equipment (1910.132)

The standards that deal with PPE consist of three types of requirements. Section 1910.132 is a set of general requirements that cover all types of equipment and situations where it is needed.

However, although no specific regulations cover the working conditions, the general requirements of 1910.132 must be met. If eye and face, respiratory, head, hand and foot protection, or equipment for electrical workers, is used as a safety precaution around any type of hazard, it must conform to 1910.133–1910.138. For a hazard regulated by any other standard, OSHA may specify protective equipment along with other required safety precautions.

OSHA does not recommend PPE if administrative or engineering controls can eliminate personal protection. Therefore, even when PPE is properly selected and used, a non-serious citation can be issued if feasible administrative or engineering controls could have been instituted.

The first step in determining what type of PPE workers need is to conduct a hazard assessment at the workplace. This is carried out by doing a walk through the work area and observing every operation in terms of the hazards it presents. Where each hazard is identified, some type of control needs to be put in place. Most times, this means PPE is necessary.

This section is designed to provide guidance on the use of PPE most commonly used for protection of the head, including eyes, ears, and the torso, arms, hands, and feet. This is intended to be guidance toward developing a policy and procedure for use of PPE. It should be noted that only some information may be applicable to all corporations and this guidance is based on industrial application.

Definitions

Personal Protective Equipment: Includes devices and clothing designed to be worn or used for the protection or safety of an individual while in potentially hazardous areas or performing potentially hazardous operations.

RESPONSIBILITIES/PROCEDURES

General Requirements

Personal protective equipment should not be used as a substitute for engineering controls or work practices. It should be used in conjunction with these controls to provide for employee safety and health in the workplace. The basic element of any management program for PPE should be an in depth evaluation of the equipment needed to protect against the hazards at the work place. Management should use that evaluation to set a standard operating procedure for personnel, then train employees on the protective limitations of personal protective equipment, and its proper use and maintenance.

Hazard Assessment

Employers are required to assess the workplace to determine whether hazards that require the use of PPE are present or are likely to be present. If hazards or the likelihood of hazards are found, the employer should select PPE suitable for protection from existing hazards.

Training

Each employee required to use PPE should be trained and able to demonstrate the ability to use PPE properly, and possibly be medically tested.

Head Protection

Prevention of head injuries is an important factor in every safety program. Head injuries are caused by falling or flying objects, such as working below other employees who are using tools and materials that could fall, or by bumping the head against a fixed object. Head protection, in the form of protective hats, must do two things: (1) resist penetration and (2) absorb the shock of a blow. This is accomplished by making the shell of the hat of a material hard enough to resist the blow, and by utilizing a shock-absorbing lining composed of headband and crown straps to keep the shell away from the wearer's skull. Protective hats are also used to protect against electrical shock.

Selection

Each type and class of head protectors is intended to provide protection against specific hazardous conditions. An understanding of these conditions will help in selecting the right hat for the particular situation.

Protective hats are made in the following types and classes:

Type 1—helmets with full brim, not less than 1 and 1/4 inches wide
Type 2—brimless helmets with a peak extending forward from the crown

For industrial purposes, three classes are recognized:

Class A—general service, limited voltage protection
Class B—utility service, high-voltage helmets
Class C—special service, no voltage protection

Class A hats and caps are intended for protection against impact hazards. They should be used in industrial facilities where potential head injury exists. Class B utility service hats and caps protect the wearer's head from impact and penetration by falling or flying objects and from high-voltage shock and burn. Class C hats and caps are designed specifically for lightweight comfort and impact protection. This class is usually manufactured from aluminum and offers no dielectric protection. Class C helmets are used where there is no danger from electrical hazards or corrosion. They are also used where there is a possibility of bumping one's head against a fixed object.

Materials used in helmets should be water-resistant. Each helmet consists essentially of a shell and suspension. Ventilation is provided by a space between the headband and the shell. Each helmet should be accompanied by instructions explaining the proper method of adjusting and replacing the suspension and headband. The wearer should be able to identify the type of helmet by looking inside the shell for the manufacturer, ANSI designation, and class.

Fit

Headbands are adjustable in 1/8-inch-size increments. When the headband is adjusted to the right size, it provides sufficient clearance between the shell and the headband. The removable or replaceable sweatband should cover at least the forehead portion of the headband. The shell should be of one-piece seamless construction and be designed to resist the impact of a blow from falling material. The internal cradle of the headband and sweatband forms the suspension. Any part that comes into contact with the wearer's head must not be irritating to normal skin.

Inspection and Maintenance

Manufacturers should be consulted with regard to paint or cleaning materials for their helmets because some paints and thinners may damage the shell and reduce protection by physically weakening it or negating electrical resistance.

A common method of cleaning shells is dipping them in hot water containing a good detergent for at least a minute. Shells should be scrubbed and rinsed in clear hot water. After rinsing, the shell should be visually inspected for any signs of dents, cracks, penetration, or other damage that might reduce the degree of safety originally provided.

Helmets should not be stored or carried on the rear-window shelf of an automobile, since sunlight and extreme heat may adversely affect the degree of protection.

Eye and Face Protection

Eye and face protective equipment is required by OSHA where there is a reasonable probability of preventing injury when such equipment is used. Employers should provide a type of protector suitable for work to be performed and employees must use the protectors. These stipulations also apply to supervisors and management personnel, and should apply to visitors while they are in hazardous areas.

Suitable eye protectors must be provided where there is a potential for injury to the eyes or face from flying particles, molten metal, liquid chemicals, acids or caustic liquids, chemical gases or vapors, potentially injurious light radiation, or a combination of these. Such equipment includes safety glasses, chemical goggles, face shields, welding goggles, and welding face shields. Protectors must meet the following minimum requirements:

- Provide adequate protection against the particular hazards for which they are designed. (Side shields for protection are highly recommended.)
- Be reasonably comfortable when worn under the designated conditions.
- Fit snugly without interfering with the movements or vision of the wearer.
- Be durable.
- Be capable of being disinfected.
- Be easily cleanable.
- Be kept clean and in good repair.
- Every protector must be distinctly marked to facilitate identification of the manufacturer only.

Each affected employee should use equipment with filter lenses that have a shade number appropriate for protection from injurious light radiation in the work being performed.

OSHA recommends that emergency eye washes be placed in all hazardous locations. First-aid instruction should be posted close to potential danger spots since any delay to immediate aid or an early mistake in dealing with an eye injury can result in lasting damage.

Selection

Each eye, face, or face-and-eye protector is designed for a particular hazard. In selecting the protector, consideration should be given to the kind and degree of hazard, and the protector should be selected on that basis. Where a choice of protectors is given and the degree of protection required is not an important issue, worker comfort may be a deciding factor.

Persons using corrective spectacles and those who are required by OSHA to wear eye protection must wear face shields, goggles, or spectacles of one of the following types:

- Spectacles with corrective lenses providing optical correction.
- Goggles worn over corrective spectacles without disturbing the adjustment of the spectacles.
- Goggles that incorporate corrective lenses mounted behind the protective lenses.
- Goggles are manufactured in several styles for specific uses such as protecting against dust and splashes, and in chipper's, welder's, and cutter's models.
- Safety spectacles require special frames. Combination of normal streetwear frames with safety lenses are not in compliance.
- Many hard hat and nonrigid helmets are designed with face- and eye-protective equipment.
- Design, construction, tests, and use of eye and face protection purchased prior to July 5, 1994, must be in accordance with ANSI Z87. 1-1968.

Fit

Fitting of goggles and safety spectacles should be done by someone skilled in the procedure. Prescription safety spectacles should be fitted only by qualified optical personnel.

Inspection and Maintenance

It is essential that the lenses of eye protectors be kept clean. Continuous vision through dirty lenses can cause eye strain—often an excuse for not wearing the eye protectors. Daily inspection and cleaning of the eye protector with soap and hot water, or with a cleaning solution and tissue, is recommended. Safety eyewear that has been previously used should be disinfected before being issued to another employee.

Ear Protection

Exposure to high noise can cause hearing impairment. It can create physical and psychological stress. There is no cure for noise-induced hearing loss, so the prevention of excessive noise exposure is the only way to avoid hearing damage. Specifically designed protection is required, depending on the type of noise encountered and the auditory condition of the employee.

Preformed or molded earplugs should be individually fitted by a professional. When properly inserted, they work as well as most molded earplugs. The individually fitted earplugs are fitted by a professional because they are only for a specific task determined with a sound level meter measured at a slow response. Due to a worker's mobility, personal monitoring is conducted by the use of a noise dosimeter.

Some earplugs are disposable, to be used one time and then thrown away. The non-disposable type should be cleaned and inspected after each use for proper protection.

Ear muffs need to make a perfect seal around the ear to be effective. Glasses, long sideburns, long hair, and facial movements such as chewing, can reduce protection. Personal protective equipment should be provided and must be used if feasible engineering controls fail to reduce sound levels to the levels specified by Title 29 CFR 1910.95—Occupational Noise Exposure.

Minimum hearing protection requirements should include the following:

Hearing protection to be available to all employees in areas where the 8-hour TWA (the allowable time-weighted average concentrations for a normal 8-hour work day or 80-hour work week) sound level equals or exceeds 85 dBA.

The employee's supervisor should enforce the wearing of hearing protection by affected employees.

Several kinds of protectors should be available to employees, thus allowing for personal preference and proper fit.

Training in the use of the hearing protectors required on an annual basis.

All persons entering posted areas are to wear hearing protection in accordance with the posted warning.

Respiratory Protection

In the control of those occupational diseases caused by articulates (dust/mist/fumes) or gases/vapors, the primary objective is to prevent atmospheric contamination. This is accomplished, as far as possible, by accepted engineering controls, such as enclosure or confinement of the operation, general and local ventilation, and substitution of less toxic materials. When effective engineering controls are not feasible, or while they are being instituted, appropriate respirators should be worn.

Requirements for a minimal acceptable program:

Employee fit-testing for respirator usage.

Employee medical evaluation.

Provide the respirator that is applicable and suitable for the purpose intended.

Establishment and maintenance of a written respiratory protection program to include the requirement outline in 29 CFR 1910.134.

The employees should use the provided respiratory protection in accordance with instructions and training received.

Respirator should be selected on the basis of hazard to which the worker is exposed.

The user should be instructed and trained in the proper use of respirators and their limitations.

Respirators should be regularly cleaned and disinfected. Those used by more than one worker should be thoroughly cleaned and disinfected after each use.

Respirator should be stored in a convenient, clean, and sanitary location.

Respirators used routinely should be inspected during cleaning. Worn or deteriorated parts must be replaced. Respirators for emergency use such as self-contained devices should be thoroughly inspected at least once a month and after each use.

Appropriate surveillance of work area condition and degree of employees exposure or stress must be maintained.

There should be regular inspections and evaluations to determine the continued effectiveness of the program.

Employees should not be assigned to tasks requiring use of respirators unless it has been determined that they are physically able to perform the work and use the equipment. A physician should determine what health and physical conditions are pertinent. The respirator user's medical status will be reviewed on an annual basis. OSHA says "periodically."

NIOSH-approved or -accepted respirators should be used when they are available. The respirator furnished should provide adequate respiratory protection against the particular hazard for which it is designed.

For more information on selection of respirator see Title 29 CFR 1910.134.

Torso Protection

Many hazards can threaten the torso: heat, splashes from hot metals and liquids, impacts, cuts, and acids. A variety of protective clothing such as vests, jackets, aprons, coveralls, and full bodysuits is available.

Selection

Duck, a closely woven fabric, is good for light-duty protective clothing. It can protect against cuts and bruises on jobs where employees handle heavy, sharp, or rough material.

Heat-resistant material, such as leather, is often used in protective clothing to guard against dry heat and flame. Rubber and rubberized fabrics, neoprene, and plastics give protection against some acids and chemicals.

Disposable suits of plastic or other similar material (Tyvek) are particularly important for protection from dusty materials or materials that can splash or spray, such as paint.

It is important to refer to the manufacturer's selection guides for the effectiveness of specific materials against specific chemicals.

Hand and Arm Protection

Many types of gloves are available to protect against a wide variety of hazards. The nature of the hazard and the operation involved will affect the selection of gloves. The variety of potential occupational hand injuries makes selecting the right pair of gloves challenging. It is essential that employees use gloves specifically designed for the hazards and tasks found in their workplace because gloves designed for one function may not protect against a different function, even though they may appear to be an appropriate protective device.

When engineering and administrative controls are not feasible or effective, PPE should be implemented. This program should address the hazards present; the selection, maintenance, and use of PPE; the training of employees; and the monitoring of the program to ensure its ongoing effectiveness.

PPE works by creating a barrier between the employee and the particular hazard he or she is dealing with. Again, it is not a substitute for good engineering or administrative controls or good work practices. However, it should be used in conjunction with these controls to ensure the safety and health of employees.

PPE must be provided, used and, maintained when it has been determined that its use is required and that such use will lessen the likelihood of occupational injury or illness.

Foot and Leg Protection

For protection of feet and legs from falling or rolling objects, sharp objects, molten metal, hot surfaces, and wet slippery surfaces, workers should use appropriate foot guards, safety shoes, or boots and leggings. Leggings protect the lower legs and feet from molten metal or welding sparks.

Aluminum alloy, fiberglass, or galvanized steel footguards can be worn over the usual work shoes. Safety shoes should be sturdy and have an impact-resistant toe. In some instances, metal insoles protect against puncture wounds. Safety shoes must conform to ANSI Z41-1991 standards. The inner lining of safety shoes are stamped with ANSI Z41 identification mark.

Training and Record Keeping

All employees using PPE should participate in a training program that must be provided at no cost to the employee and during working hours.

Training is to be as follows:

At the time of initial assignment to a task where PPE is used.

Retraining should be required when changes in the workplace or types of PPE to be used render previous training obsolete, or if inadequacies in an employee's knowledge or use of assigned PPE indicate that the employee has not retained the requisite understanding or skill.

Material appropriate in content and vocabulary to education level, literacy, and language of employees should be used.

The training program should contain at a minimum the following elements.

When PPE is necessary, what PPE is necessary, how to don, doff, adjust, and wear PPE, limitations of the PPE, and proper care, maintenance, useful life, and disposal of the PPE.
Information on the types, proper use, location, removal, handling, decontamination, and disposal of PPE.
An explanation of the procedure to follow if an incident occurs, including the method of reporting the incident and the medical follow-up that should be made available.
An opportunity for interactive questions and answers with the person conducting the training session, who shall be knowledgeable in the subject matter covered by the elements contained.

Training records shall include the following:

Names of training attendants.
The dates of the training sessions.
The contents or a summary of the training sessions.
The names and qualifications of persons conducting the training sessions.
Training records shall be maintained for at least 3 years from the date on which the training occurred.

REVIEW QUESTIONS

TRUE OR FALSE

1. T or F—Protective gloves should be inspected before each use to ensure that they are not torn, punctured, or made ineffective in any way.
2. T or F—The OSHA standards dealing with PPE consist of three types of requirements.
3. T or F—Always conduct a workplace hazard assessment to reveal that employees face potential injury that cannot be eliminated through engineering and work practice controls.
4. T or F—Prevention of head injuries is not an important factor in safety programs.
5. T or F—Hard hats consist essentially of a shell and suspension.
6. T or F—Face and eye protection does not have to be designed for a particular hazard.
7. T or F—Exposure to high noise can cause hearing loss.
8. T or F—Hearing protection is to be available to all employees in areas where the 8-hour TWA sound level is below 85 dBA.
9. T or F—Respirators shall be stored in a convenient, clean, and sanitary location.
10. T or F—Training in the use of the hearing protectors is required every 5 years.

11. Training records shall be maintained for at least ______________ years from the date on which the training occurred.
12. ______________ restricted material, such as leather, is often used in protective clothing.
13. ______________ approved or accepted respirators should be used when they are available.
14. ______ ______ ______ should not be used as a substitute for engineering controls or work practices.
15. OSHA recommends that emergency ______________ watches be placed in all hazardous locations.

BIBLIOGRAPHY

29 CFR 1910 Subpart I Personal Protective Equipment
CFR 1910.132 General requirements
CFR 1910.133 Eye and face protection
CFR 1910.134 Respiratory protection
CFR 1910.135 Occupational head protection
CFR 1910.136 Occupational foot protection
ANSI Z87.1-1968 Eye and face protection
ANSI Z1-1991 Men's Safety-toe Footwear
Kohn, James P., Friend, Mark A. and Winterberger, Celeste A. Fundamentals of occupational safety and health, Government Institutes, 1996.
Germain, George L., Arnold, Robert M., Rowan, J. Richard and Roane, J. R. Safety, health and environmental management, AEI and Associates, 1997.
The occupational environment: Its evaluation and control, American Industrial Hygiene Association, 1997.
Respiratory protection OSHA 3079, 1998.

Note: All CFR codes can be obtained by contacting the Office of Occupational Safety and Health Standards, Washington, D.C. (U. S. Government Printing Office).

11 Noise and Ventilation

INTRODUCTION

This chapter deals with occupational noise and ventilation procedures.

All employees exposed at or above an 8-hour TWA of 85 dBA (action level) or more are covered by this regulation (OSHA-1910.95—Noise Exposure). When you include all human senses, only vision has a higher rate of information transfer. We know that sound is the propagation, transmission, and reception of waves in some medium, usually air. Sound waves may have a single frequency or might be a combination of frequencies. Short-duration noise pulses, which can occur one time or may be repetitive, are known as impulse noise.

The many effects of noise on individuals can lead to several kinds of hearing loss. High frequencies are more damaging when compared with low frequencies, supporting the fact that a continual noise is more of a problem that intermittent noise problems.

The purpose of the noise section is to provide protection against the effects of noise exposure through noise monitoring, audiometric testing, determination of threshold shifts, hearing protection, employee training, and record keeping. Determination of the dBA should be computed in adherence to 1910.95 Appendix A—Noise Exposure Computation.

Definitions

Abrasive: A solid substance used in an abrasive blasting operation.

Abrasive Blasting Respirator: A continuous flow air-line respirator constructed so that it will cover the head, neck, and shoulders to protect from rebounding abrasive.

Blast Cleaning Barrel: A complete enclosure that rotates on an axis, or has an internal moving tread to tumble the parts in order to expose various surfaces of the parts to the action of an automatic blast spray.

Blast Cleaning Room: A complete enclosure in which blasting operations are performed and where the operator works inside of the room to operate the basting nozzle and direct the flow of the abrasive material.

Blasting Cabinet: An enclosure where the operator stands outside and operates the blasting nozzle through an opening or openings in the enclosure.

Clean air: Air of such purity that it will not cause harm or discomfort to an individual if it is inhaled for extended periods of time.

Dust Collector: A device or combination of devices for separating dust from the air handled by an exhaust ventilation system.

Exhaust Ventilation System: A system for removing contaminated air from a space, comprising two or more of the following elements (a) enclosure or hood (b) duct work (c) dust collecting equipment (d) exhauster and (e) discharge stack.

Particulate Filter Respirator: An air purifying respirator, commonly referred to as a dust or fume respirator, which removes most of the dust or fume from the air passing through the device.

Respirable Dust: Airborne dust in sizes capable of passing through the upper respiratory system to reach the lower lung passages.

Rotary Blast Cleaning Table: An enclosure where the pieces to be cleaned are positioned on a rotating table and are passed automatically through a series of blast sprays.

Abrasive Blasting: The forcible application of an abrasive to a surface by pneumatic pressure, hydraulic pressure, or centrifugal force.

Grinding, Polishing and Buffing Operations

Abrasive cutting-off wheels: Organic-bonded wheels the thickness of which is not more than one 48th of their diameter for those up to and including 20 inches in diameter and not more than one 60th of their diameter for those larger than 20 inches in diameter, used for a multitude of operations variously known as cutting, cutting off, grooving, slotting, coping, jointing, and the like.

Belts: All power-driven, flexible coated bands used for grinding, polishing, or buffing purposes.

Branch Pipe: The part of an exhaust system piping that is connected directly to the hood or enclosure.

Cradle: A movable fixture upon which the part to be ground or polished is placed.

Disc Wheels: All power-driven rotatable discs faced with abrasive material, artificial or natural, and used for grinding or polishing on the side of the assembled disc.

Entry Loss: The loss in static pressure caused by air flowing into a duct or hood. It is usually expressed in inches of water gauge.

Exhaust System: A system consisting of branch pipes connected to hoods or enclosures, one or more header pipes, an exhaust fan, means for separating solid contaminants from the air flowing in the system, and a discharge stack outside.

Grinding Wheels: All power-driven rotatable grinding or abrasive wheels except disc wheels as defined in this standard consisting of abrasive particles held together by artificial or natural bonds and used for peripheral grinding.

Header Pipe (Main Pipe): A pipe into which one or more branch pipes enter and that connects such branch pipes to the remainder of the exhaust system.

Hoods and Enclosures: The partial or complete enclosure around the wheel or disc through which air enters an exhaust system during operation.

Horizontal Single-Double Disc Grinder: A grinding machine carrying two power-driven, rotatable, coaxial horizontal spindles upon the inside ends of which are mounted abrasive disc wheels used for grinding two surfaces simultaneously.

Horizontal Single-Spindle Disc Grinder: A grinding machine carrying an abrasive disc wheel on one or both ends of a power-driven, rotatable single horizontal spindle.

Polishing and Buffing Wheels: All power-driven rotatable wheels composed all or in part of textile fabrics, wood, felt, leather, paper. May be coated with abrasives on the periphery of the wheel for purposes of polishing, buffing, and light grinding.

Portable Grinder: Any power-driven rotatable grinding, polishing, or buffing wheel mounted in such a manner that it can be manually manipulated.

Scratch Brush Wheels: All power-driven rotatable wheels made from wire or bristles, and used for scratch cleaning and brushing purposes.

Swing-Frame Grinder: Any power-driven rotatable grinding, polishing, or buffing wheel mounted on such a manner that the wheel with its supporting framework can be manipulated over stationary objects.

Velocity Pressure: The kinetic pressure in the direction of flow necessary to cause a fluid at rest to flow at a given velocity. It is usually expressed in inches of water gauge.

Vertical Spindle Disc Grinder: A grinding machine having a vertical, rotatable power-driven spindle carrying a horizontal abrasive disc wheel.

Spray-Finishing Operations

Spray-Finishing Operations: Are employment of methods wherein organic or inorganic materials are utilized in dispersed form for deposit on surfaces to be coated, treated, or cleaned. Such methods of deposit may involve either automatic, manual, or electrostatic deposition but do not include metal spraying or metalizing, dipping, flow coating, roller coating, tumbling, centrifuging, or spray washing and degreasing as conducted in self-contained washing and degreasing machines or systems.

Spray Booth: A power-ventilated structure provided to enclose or accommodate a spraying operation to confine or limit the escape of spray, vapor, and residue, and to safely conduct or direct them to an exhaust system.

Spray Room: A room in which spray-finishing operations not conducted in a spray booth are performed separately from other areas.

Minimum Maintained Velocity: The velocity of air movement that must be maintained to meet minimum specified requirements for health and safety.

Occupational Noise

Action Level: Level at which action must be taken to protect the worker from a hazard.

Audiogram: A record of an individual's hearing loss at hearing level measured at several different frequencies presented numerically or graphically.

Audiologist: A person with graduate training in the specialized problems of hearing and deafness.

Baseline Audiogram: The audiogram against which future audiograms are compared.

Criterion Sound Level: A sound level of 90 decibels.

Decibel (dB): Unit of measurement of sound level.

Noise Dosimeter: An instrument that integrates a function of sound pressure over a period of time in such a manner that it directly indicates a noise dose.

Representative Exposure: Measurements of an employee's noise dose or 8-hour TWA sound level that the employers deem to be representative of the exposures of other employees in the workplace.

Sound Level: Measurement of the A-weighted sound pressure indicated in dBA.

RESPONSIBILITIES/PROCEDURES

Responsibilities

The *corporation* is responsible for issuing the policy on workplace ventilation requirements to the operating facility and updating the policy as necessary to ensure compliance with the most current regulations. The *operating facilities safety department* is responsible for administering this policy, conducting any workplace monitoring that is necessary to fulfill requirements and enforcing its use. The *operations manager* is responsible for implementing the policy, communicating it to the workforce and enforcing its use. Each *employee* is responsible for reporting any problems to his or her *supervisor* who, in turn, is responsible for evaluation and correction.

Procedures

Abrasive Blasting

The concentration of respirable dust or fumes in the breathing zone of any worker shall be kept below the levels specified in Sec. 1910.94 of the OSHA standard.

Combustible organic abrasives are to be used only in automatic systems. The exhaust system, and all electrical wiring where explosive or flammable dust mixtures are present, must meet the requirements of ANSI Z33. 1-1961 Installation of Blower and Exhaust Systems for Dust, Stock and Vapor removal or Conveying, and NFPA 70-1971 National Electrical Code. The blast nozzle should be bonded and grounded. The abrasive blasting enclosure ducts and dust collector must be constructed with loose panels or explosion venting areas located on sides away from any occupied areas in accordance with NFPA 68-1954 National Fire Protection Association Explosion Venting Guide.

Blast cleaning enclosures shall provide a continuous inward flow of air at all openings during the blasting operation.

Air inlets and openings shall be arranged so as to minimize escape of abrasive and dust particles into the adjacent work areas.

The rate of exhaust shall be sufficient to provide prompt clearance of dust-laden air within the enclosures after cessation of blasting

The exhaust system shall run a sufficient time after the blast is turned off to remove dusty air from within the enclosure.

Safety glass shall be protected by screening use in observation windows where hard, deep-cutting abrasives are used.

Slit abrasive-resistant baffles shall be installed in multiple sets at all small access openings where dust might escape. They shall be inspected regularly and replaced when needed.

Doors shall be flanged and tight when closed.

Doors on blast-cleaning rooms shall be operable from both inside and outside.

The construction, installation, inspection, and maintenance of the exhaust system shall comply with the requirements of ANSI Z9.2-1960, Fundamentals Governing the Design and Operation of Local Exhaust Systems.

Dust leaks when noted, shall be repaired as soon as possible.

The static pressure drops at exhaust ducts shall be checked periodically and the system cleaned when a pressure drop indicates a partial blockage.

An abrasive separator shall be provided where the abrasive is recirculated.

The exhausted air from blast cleaning equipment shall be discharged through dust collecting equipment that can be emptied and removed without contaminating working areas.

NIOSH-approved respiratory equipment shall be used for protection against dust produced during abrasive-blasting operations.

All abrasive-blasting operators shall wear approved respirators when:

- Working inside blast-cleaning rooms.
- Using silica sand in manual blasting operations where the nozzle and blast aren't physically separated from the operator in an exhaust ventilated enclosure.
- Concentrations of toxic dust exceed limits fixed by OSHA standards, and nozzle and blast aren't physically separated from the operator in an exhaust ventilated enclosure.

Dust-filter respirators shall be used for short intermittent exposures such as cleanup, dumping of dust collectors, unloading, etc., where exhaust ventilation isn't feasible.

The respiratory protection program shall comply with Sec. 1910. 134(a-b) of the OSHA Standards.

Operators shall be equipped with heavy canvas or leather gloves and aprons. Safety shoes shall be worn when heavy pieces of work are handled. Eye and face protection shall be supplied where the respirator does not afford sufficient protection.

Air from the abrasive-blasting respirators shall meet the requirements of ANSI Z9.2-1960, Fundamentals Governing the Design and Operation of Local Exhaust Systems. Where air from the regular compressed air line is used there shall be:

A trap or filter to remove oil, water, scale, and odor

A pressure reducing valve

An automatic control to warn in case of overheating

All dust shall be cleaned up promptly. Aisles and walkways shall be kept clear of abrasives that might cause a slipping hazard.

Grinding, Polishing, and Buffing Operations

Any employees engaged in dry grinding, dry polishing, or buffing exposed beyond the limits specified in Sec. 1910.1000 of the OSHA Standards shall be provided with a local exhaust ventilation system to maintain employee exposure within prescribed limits.

Hoods shall be connected to exhaust systems and shall be designed, located, and placed so that dust is projected into the hoods in the direction of air flow.

The minimum exhaust volume shall not be less than that specified in the OSHA standards.

All exhaust systems shall be provided with suitable dust collectors and they shall be designed and tested in accordance with ANSI B7.1-1960, Fundamentals Governing the Design and Operation of Local Exhaust Systems.

The structural strength of the hoods shall be sufficient to protect the operator from the hazards of bursting wheels. They shall meet the strength requirements of ANSI B7.1-1970 Safety Code for the use, Care and Protection of Abrasive Wheels.

The hoods shall be located as close as possible to the operation.

Exhaust hoods for floor stands, pedestals, and bench grinders shall be designed in accordance with OSHA standards.

Exhaust booths for swing-frame grinders shall comply with OSHA Standards.

Portable grinding operations shall be conducted within a partial enclosure not larger than actually required and with an average face air velocity of at least 200 feet per minute.

Hoods and enclosures shall be designed in accordance with OSHA standards.

Spray-Finishing Operations

Spray booths or spray rooms shall be used to confine all spray-finishing operations.

Spray booths shall be designed and constructed in accordance with OSHA Standards and NFPA No.33-1969, Standard for Spray Finishing Using Flammable and Combustible Materials.

Unobstructed walkways shall be at least 6 1/2 feet high and clear of obstructions from any work location to an exit. Single exits shall be at least 3 feet wide. Multiple exits may be 2 feet wide if the maximum distance from work locations is less than 25 feet. Exit doors must swing outward.

Baffles, distribution plates, and dry-type overspray collectors shall conform to OSHA standards. Overspray filters (where used) shall be installed and maintained in accordance with OSHA standards.

Where the booth is made of steel, the water chamber enclosure for wet or water-wash spray booths shall be 18 gauge or heavier and protect against corrosion. The chamber shall effectively remove particulate matter from the exhaust air stream.

Collecting tanks shall be made of suitable noncombustible material. They should be designed to prevent sludge and floating paint from entering the pump suction box. The proper water level shall be automatically maintained. Fresh water inlets shall not be submerged. Precautions shall be taken not to accumulate hazardous deposits in the tanks.

Sufficient water flow shall be insured to provide efficient operation of the water chamber.

Spray rooms, including floors, shall be constructed of non-combustible materials.

Spray rooms shall have non-combustible fire doors and shutters.

Ventilation shall be provided in accordance with OSHA Standards and NFPA 33-1969. Fans shall be adequate to control distribution of exhaust air movement through the booth.

Inlet and supply ductwork shall be made of non-combustible materials. Seams and joints of inlet ductwork shall be sealed. Inlet ductwork shall be sized in accordance with volume flow requirements. Inlet ductwork shall be adequately supported throughout its length.

Ducts shall be constructed of adequate gauge steel with a diameter or greater dimension based on U.S. gauge:

Up to 8 in. inclusive	No. 24
Over 8 in. to 18 in. inclusive	No. 22
Over 18 in. to 30 in. inclusive	No. 20
Over 30 in.	No. 18

Exhaust ductwork shall be adequately supported and sized.

Longitudinal joints in sheet steel ductwork shall be lock-seamed, riveted, or welded. Circumferential joints shall be fastened and lapped in the direction of the airflow. At least every fourth joint shall be flanged or bolted.

Inspection or clean-out doors shall be provided for every 9 to 12 feet of running length for ducts up to 12 in. diameter. A clean-out door shall be provided for servicing the fan. A drain shall be provided where necessary.

Protection shall be provided where ductwork passes through a combustible roof or wall. Automatic fire dampers or steel plates shall be provided where ductwork passes through firewalls.

Ductwork used for ventilating spray finishing operations shall not be connected to ducts used for ventilating other processes, to chimneys or flues used for combustion.
The velocity of air in all openings of the spray booth shall not be less than specified in Sec. 1910.94(c)(6)(i) of the OSHA standards.
Operators in booths downstream of the object being sprayed shall be required to wear an approved respirator.
Downdraft booth doors shall be closed when spray painting.
Clean fresh air shall be supplied to spray booths or rooms equal in quantity to the volume of air exhausted.
Doors, dampers, or louvers through which make-up air is supplied shall be open when the booth or room is being used for spraying. The air velocity shall not be in excess of 200 feet per minute.
Filters through which make-up air is supplied shall have a gauge marked to show the pressure drop at which the filters require cleaning or replacement. Filters shall be replaced or cleaned when air flow drops below specified levels.
When outside temperature is below 55°F and radiant heating isn't provided, makeup air shall be heated to not less than 65°F at the point of entry. (General heating of the building to at least 65°F during exhaust operation shall be a suitable alternative)
Precautions shall be taken to insure that the heating system for the make-up air is not located in the spray booth.

Reports, Forms, and Record Keeping

Inspection and maintenance records of all local exhaust systems should be kept in accordance with the requirements of ANSI Z9.2-1960.

Items to be included are:

dates of filter changes
service dates of fans and blowers
air quality testing records
results of the tests of any exhaust systems
repair records

Training/Certification

All operators of grinding, blasting, and spray-finishing operations shall be provided training on respirators appropriate for the jobs they conduct. This respiratory protection program should comply with 1910. 134(a-b). Employees should also be trained on other PPE including gloves, eye and face protection, and safety shoes.

Occupational Noise

Responsibilities

The corporation is responsible for issuing this policy (Safety Department), updating it as necessary to ensure compliance with the most recent regulations, and communicating this policy to operations management.

The safety department is responsible for administering this program, including determining facility noise levels as necessary, ensuring that employee audiometric testing is performed and appropriate records maintained as required by applicable standards, coordinating employee training programs, providing appropriate types of hearing protective devices as determined by facility noise levels and advising employees exposed at 85 dB or greater of the results of noise monitoring. The facility safety department shall undertake the implementation of any feasible engineering noise controls.

Employees are responsible for wearing hearing protective devices as established by facility noise monitoring results, and reporting any problems with the use of hearing protective devices to appropriate supervision. Employees shall also be responsible for reporting for audiometric testing when notified.

Procedures

The noise exposure of employees shall be determined by measuring the sound levels with a sound level meter measured on the A scale at slow response. Where area monitoring is not feasible (due to worker mobility, etc.) personal monitoring by use of a noise dosimeter shall be conducted.

Protection against the effects of noise exposure shall be provided when sound levels exceed the following when measured on the A scale of a standard sound level meter at slow response:

Duration per Day

Number of Hours	Sound Level at dBA
8	90
6	92
4	95
3	97
2	100
1.5	102
1	105
.5	110
.25 or less	115

Where employees are exposed to noise exceeding the permissible exposure limits, engineering controls shall be utilized. If this is not feasible, administrative controls or PPE shall be utilized to reduce sounds to the levels within the regulations.

Employee noise exposure shall be computed in accordance with the regulation and without regard to any reduction provided by the use of PPE.

Where an employee's exposure equals or exceeds an 8-hour TWA of 85 dB (action level) a monitoring program shall be developed and implemented.

All employees exposed to noise levels shall be identified for inclusion in the hearing conservation program and to enable the proper selection of hearing protectors.

Continuous, intermittent, and impulsive sound levels from 80 to 130 dB shall be integrated into the noise measurements.

Monitoring shall be repeated whenever a change in production, process, equipment, or controls increase noise exposures and additional employees may be exposed or the reduction of noise exposure by personal protective equipment is rendered inadequate.

Employees exposed at or above the action level will be notified of the results of the monitoring and employees or their representatives will be permitted the opportunity to observe any noise measurements.

An audiometric testing program shall be established and maintained at no cost to employees whose exposure equals or exceeds an 8-hour TWA of 85 decibels.

These tests shall be performed by a licensed or certified audiologist, otolaryngologist, or other physician certified in occupational hearing conservation or who has demonstrated competence in administering audiometric examinations.

A baseline audiogram shall be established on employees exposed at or above the action level within 6 months of the first exposure. Employees shall be notified of the requirement that testing shall be preceded by at least 14 hours without exposure to workplace noise. Hearing protectors may be worn as a substitute to this requirement.

An annual audiogram shall be obtained on employees at or above the 8-hour TWA of 85 dB and compared annually with the baseline audiogram. If a threshold shift has occurred, the audiogram will be evaluated and the employee informed in writing.

Threshold shifts that are work related or aggravated by occupational noise exposure shall require training and fitting on hearing protection or retraining on hearing protection, with concomitant increase in the offered hearing protection.

Hearing protectors shall be made available to employees exposed to an 8-hour TWA of 85 dB or greater at no cost to the employees, replaced as necessary and their use enforced. Employees shall be allowed to choose from a variety of suitable hearing protectors.

Training shall be provided on the use and care of all hearing protectors provided to employees.

Employees shall undergo an initial fitting and be supervised on the correct use of all hearing protectors.

Hearing protection shall be evaluated to ensure that it reduces the employee noise exposure as required by the regulation. Re-evaluation will be conducted where increases in noise exposures may render previous hearing protection inadequate.

Reports, Forms, and Record Keeping

The company shall maintain records of employee exposure measurements 30 years or 10 years beyond the employment of the employee, whichever is later.

The company shall retain all employee audiometric testing records pursuant to the audiometric testing program requirements of this regulation.

These records shall be retained as follows:

Noise exposure records = 2 years

Audiometric test records = duration of employee's employment

These records are available upon request to employees, former employees, representatives of the employee, and the assistant secretary.

These records shall be transferred to any company successor who shall be required to maintain them in accordance with the regulatory requirements.

Copies of the OSHA Occupational Noise Exposure standard (1910.95) shall be provided to employees or their representatives and a copy shall be posted in the workplace in a location that is easily accessible to employees.

Any additional material relating to this standard as supplied by the assistant secretary shall also be provided.

Training/Certification

A training program shall be instituted for all employees exposed at or above an 8-hour TWA of 85 dB and participation in such program is mandatory.

This training program shall be repeated annually for all employees included in the hearing conservation program. The training information shall be updated to remain consistent with changes in protective equipment and work processes.

Training material shall include:

Effects of noise on hearing

Purpose of hearing protection

Instructions on selection, fit, use and care

Purpose of audiometric testing

Explanation of audiometric test procedures

Additional training sessions/materials on hearing conservation may be provided to the employees through publications, employee mailings, safety meetings, or other forums as determined by the facility's safety department.

REVIEW QUESTIONS

True or False

1. T or F—Sound waves may have a single frequency or a combination of frequencies.
2. T or F—Decibel (dB) is a unit of measurement of sound levels.
3. T or F—OSHA's hearing conservation program does not include all employees exposed to an eight-hour time weighted average sound level of 85 decibels.
4. T or F—Blast cleaning enclosures will provide a continuous inward flow of air at all openings during the blasting operation.

5. T or F—It is not a requirement for spray operations to wear an approved respirator in booths that are downstream of the object.
6. T or F—All operators of grinding, blasting, and spray finishing shall be provided training on respirators appropriate for the jobs they conduct.
7. T or F—When an exposure equals or exceeds 8-hour TWA 85 dB a monitoring program shall be developed and implemented.
8. T or F—Sound waves only have one frequency and not a combination of frequencies.
9. T or F—There are four types of ventilation.
10. T or F—Training programs shall be repeated annually for all employees included in the hearing conservation program.

REFERENCES

Berger, Elliott H., Royster, Julia D., Royster, Larry, Driscoll, Dennis P., and Layne, Martha. *The noise manual, 5th ed.,* American Industrial Hygiene Association, Fairfax, VA, 2003.

Industrial hygiene OSHA 3143, 1998.

Knowles, Emory E. III, Ed. *Noise control, 3rd ed.*, American Society of Safety Engineers, Des Plaines, IL, 2003.

Dinardi, S.R., Ed. *The occupational environment: Its evaluation and control*, American Industrial Hygiene Association, 1997.

Occupational noise exposure, NIOSH Publication 98-126.

Respiratory protection, OSHA 3079, 1998.

12 Bloodborne Pathogen Standard

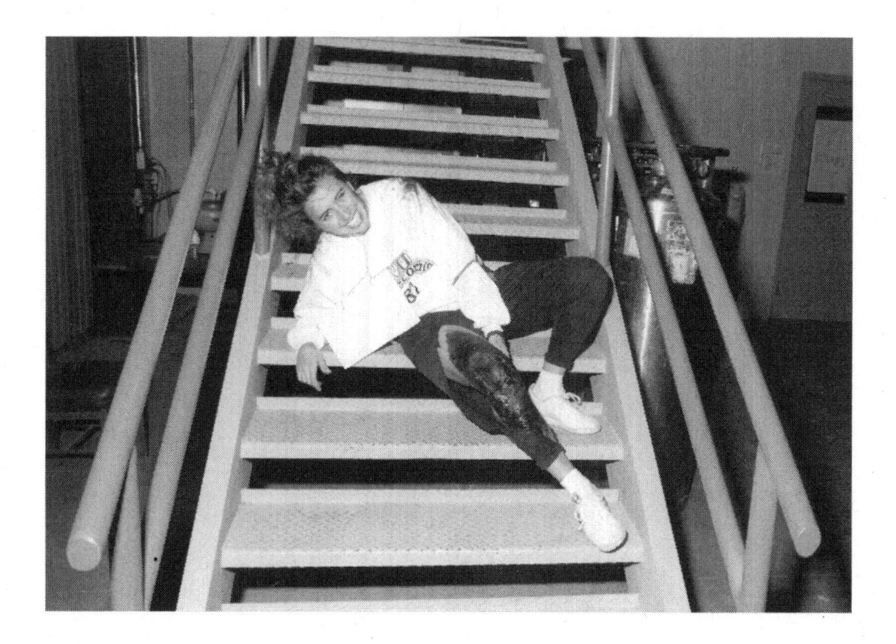

INTRODUCTION

This section is designed to protect employees whose job places them at risk of exposure to blood or other potentially infectious material. This is intended to be guidance toward developing a policy and procedure in the event of an exposure of an employee to blood or other body fluids. It should be noted that only some information may be applicable to all corporations and this guidance is based on industrial application.

In accordance with the bloodborne pathogen standard, OSHA has responded to a number of incidents where healthcare workers have become infected with human immunodeficiency virus (HIV) or hepatitis B virus (HBV) due to occupational exposures. The standard covers a relatively limited number of workers, most of them from the healthcare industries.

OSHA revised the standard in November 2000 in accordance with the Needle Stick Safety and Protection Act. The revision clarified the requirements for employers to select safer needle devices as they become available and in identifying and selecting the devices. The employer must establish and keep a sharps injury log that became effective April 18, 2001.

Definitions

Bloodborne Pathogen: Pathogenic microorganisms that are present in human blood and can cause disease in humans. These pathogens include, but are not limited to, hepatitis B virus (HBV) and human immunodeficiency virus (HIV).

Disinfectant: Agent used to decontaminate equipment or surfaces. Such disinfectants would include a 1:100 bleach solution for routine decontamination and a 1:10 solution for higher biological loads such as a blood spill caused by serious injury.

Decontamination: The use of physical or chemical processes to remove, inactivate, or destroy living organisms to some lower level (not necessarily zero).

Engineering Controls: Controls that isolate or remove the agent from the workplace.

Handwashing Facilities: A facility providing an adequate supply of running potable water, soap, and single-use towels or hot air drying machines.

Human Immunodeficiency Virus: A bloodborne pathogen that attacks the body's immune system, causing the disease known as AIDS.

Personal Protective Equipment: Specialized clothing or equipment worn by an employee for protection against a hazard.

Universal Precautions: Guidelines that require all blood and certain body fluids to be treated as if they were known to be infected with HIV, HBV, or other bloodborne diseases.

Responsibilities/Procedures

Responsibilities are to be determined by the corporation. An example for responsibilities may be stated as the following:

The safety director of the individual companies has the authority to establish a procedure to comply with the Bloodborne Pathogen Standard 29 CFR 1910.1030.

The managers and foremen are responsible for assuring that all personnel involved adhere to this procedure.

Employees are responsible for complying with the procedures involving bloodborne pathogens and proper first aid procedures in compliance with universal precautions.

Procedures

OSHA guidelines require that each employer who has employee(s) with potential occupational exposure shall prepare an exposure determination. This exposure determination shall contain the following:

A list of job classifications for all employees whose job classifications have occupational exposure.

A list of classifications in which some employees have occupational exposure.

A list of tasks and procedures (i.e., first aid procedures) or groups of closely related tasks and procedures in which occupational exposure occurs.
The schedule and method of implementation, method of compliance, hepatitis vaccination, post-exposure evaluation and follow-up, communication of hazards and record keeping required by 29 CFR 1910.1904 and 1030.
The procedure for the evaluation of circumstances surrounding incidents.

Exposure Control Plan

Job classifications in which all employees in those classifications have potential occupational exposure require an exposure-control plan developed by the Safety Department. Examples may include:

First aid responders to injuries or accidents. (This may be shop/maintenance personnel only.) Designated first aid responders established by management.
Job classifications in which some employees have some occupational exposure. Examples may include:
Supervisors
Administrative personnel
Plant security
Human resources personnel
Tasks and procedures or group of closely related tasks in which occupational exposure could potentially occur. Examples may include:

- CPR
- Automated external defibrillator (AED) training (3 hours)
- Basic first aid
- Accident response
- Clean-up of biohazadarous spills (blood) from an injury

Each year, approximately 465,000 people in the United States die from sudden cardiac arrest (SCA). The American Heart Association (AHA) recommends defibrillation response within 5 minutes for out-of-hospital events, and 3 minutes for in-hospital events. The AED was invented specifically to address the third and weakest link in the chain, early defibrillation.

Method of Compliance

General universal precautions should be observed to prevent contact with blood or other potentially infectious materials. Under circumstances in which differentiation between body fluid types is difficult or impossible, all body fluid should be considered infectious material.

Engineering and Work Practice Controls

Engineering and work practice controls should be used to eliminate exposure. Where occupational exposure remains after institution of these controls, PPE shall be used.

Personal protective equipment should be examined and maintained or replaced on a regular schedule to ensure its effectiveness. This schedule should be posted and documented.

The employer should provide handwashing facilities that are readily accessible to employees.

The employer should ensure that employees wash their hands immediately of soon as feasible after removal of gloves or PPE. Where handwashing facilities are not available, the employer should provide an antiseptic hand cleanser or antiseptic towelettes. Use these as temporary measures only.

The employer shall ensure that employees wash their hands and any other exposed skin with soap and water, or flush mucous membranes with water immediately or as soon as feasible following contact of such body areas with blood or other potentially infectious materials.

Eating, drinking, smoking, applying cosmetics, applying lip balm, or handling contact lenses should be prohibited in designated first-aid areas.

Personal Protective Equipment

When there is an potential occupational exposure, the employer should provide, at no cost to employee, appropriate personal protective equipment such as gloves, eye protection, pocket masks, or other ventilation devices. Personal protective equipment shall be considered "appropriate" only if it does not permit blood or other potentially infectious material to pass through to, or reach employee's clothes, skin, eyes, mouth or other mucous membranes under normal conditions of use.

The employer should ensure that appropriate protective equipment in the appropriate sizes is readily accessible at the worksite or issued to employees. Hypoallergeric gloves or alternatives should be accessible to those employees who are allergic to the gloves normally provided.

The employer should clean, launder, and dispose of PPE.

All PPE shall be removed prior to leaving the facility.

When PPE is removed, it shall be placed in an appropriately designated area or container for storage, washing, decontamination, or disposal.

Gloves should be worn when it can be reasonably anticipated that the employees may have contact with blood or other potentially infectious materials, mucous membranes, non-intact skin, and when handling or touching contaminated items or surfaces.

Disposable (single-use) gloves should be replaced as soon as feasible if they tear, are punctured, or when their ability to function as a barrier is compromised.

Disposable gloves shall not be washed or decontaminated for reuse.

General Housekeeping

The employer shall ensure that the worksite is maintained in a clean and sanitary condition. An appropriate schedule for cleaning and method of decontamination based on the location within the facility, type of surface to be cleaned, type of soil present, and tasks or procedures being performed in the area.

In the event of an accident where blood or other potentially infectious materials are found on equipment or surfaces, these items and surfaces should be cleaned and decontaminated.

Any bins, pails, cans, and similar receptacles used for decontamination and disposal should also be decontaminated.

Broken glass or anything that can pierce, puncture, or cut the skin that may have been contaminated should not be picked up directly with the hands. It should be gathered using mechanical means, such as a brush and dustpan, tongs, or forceps.

If waste containing blood or other potentially infectious materials is generated it is considered regulated waste. Regulated waste should be discarded immediately or as soon as feasible in containers that are closable, puncture resistant, leakproof on sides and bottom, and labeled indicating medical waste with a biohazard sign.

Hepatitis B Vaccination and Post-Exposure Evaluation and Follow-up

General Guidelines

The employer should make available Hepatitis B vaccine and vaccination series to all employees who have occupational exposure, and post-exposure evaluation and follow-up to all employees who have had an exposure incident.

The employer should ensure that all medical evaluations and procedures including the Hepatitis B vaccine and vaccination series, and post-exposure evaluation and follow-up, including prophylaxis, are:

Made available at no cost to the employee.

Made available to the employee at a reasonable time and place.

Performed by or under the supervision of a licensed physician or under the supervision of another licensed healthcare professional.

Provided according to U.S. Public Health Service recommendations current at the time these evaluations and procedures take place.

Post-Evaluation and Follow-Up

Following a report of an exposure incident the employer should immediately make available to the exposed employee a confidential medical evaluation and follow-up, including at least the following elements:

Documentation of the route(s) of exposure(s), and the circumstances under which the exposure incident occurred.

Identification and documentation of the source individual, unless the employer can establish that the identification is unfeasible or prohibited by state or local law.

The source individual's blood shall be tested as soon as feasible and after consent is obtained in order to determine HBV and HIV infectivity. If consent is not obtained, the employer shall establish that legally required consent cannot be obtained. When the source individual's consent is not required

by law, the source individual's blood, if available, shall be tested and the results documented.
Counseling.
Evaluation of reported illness.
Post-exposure prophylaxis, when medically indicated, as recommended by the U.S. Public Health Service.

When the source individual is already known to be infected with HBV or HIV, testing for the source individual's known HBV or HIV status need not be repeated.

Results of the source individual's testing shall be made available to the exposed employee, and the employee shall be informed of applicable laws and regulations concerning disclosure of the identity and infectious status of the source individual.

The employer should ensure that the healthcare professional evaluating an employee after an exposure incident is provided the following information:

A copy of 29 CFR 1910.1030.
A description of the exposed employee's duties as they relate to the exposure incident.
Documentation of the route(s) of exposure and circumstances under which exposure occurred.
Results of the source individual's blood testing, if available. All medical records relevant to the appropriate treatment of the employee, including vaccination status, which are the employer's responsibility to maintain.

Reports, Forms and Record Keeping

Employers also must preserve and maintain for each employee an accurate record of occupational exposure according to OSHA's rule governing access to employee exposure and medical records.

Under the bloodborne pathogens standards, however, medical records also must include the following information:

Employee's name and social security number
Employee's hepatitis B vaccination status
Results of examinations
Health care professional's written opinion
A copy of the information provided to the health care professional

Medical records must be kept confidential and maintained for at least the duration of employment plus 30 years.

The bloodborne pathogens standard also requires employers to maintain and keep accurate training records for 3 years and to include the following:

Training Dates
Content or a summary of the training
Names and qualifications of trainers

Names and job titles of trainees
Communication of hazards to employees
Labels and Signs (BIOHAZARD)

Warning labels should be affixed to containers of regulated waste. These labels should be fluorescent orange or orange or predominately so, with lettering of symbols in contrasting color. Red bag or red containers may be substituted for labels.

Information and Training

The company should ensure that all employees with a potential occupational exposure participate in a training program that must be provided during working hours at no cost to the employee.

Training shall be as follows:

At the time of initial assignment to task where a potential occupational exposure may take place.
Annual training for all employees should be provided within 1 year of their previous training.
The company should provide additional training when changes occur such as modification of tasks or procedure and the effect on the employee's exposure. New training may be limited to addressing the new potential exposure created.
Material appropriate in content and vocabulary to education level, literacy, and language of employees should be used.

The training program should contain at a minimum the following elements.

An accessible copy of the text of 29 CFR 1910.1030 and an explanation of its contents.
A general explanation of epidemiology and symptoms of bloodborne disease.
An explanation of the modes of transmission of bloodborne pathogens.
An explanation of this employer's exposure control plan and the means by which the employee can obtain a copy of the written plan.
An explanation of the appropriate methods for recognizing tasks and other activities that may involve exposure to blood and other potentially infectious material.
An explanation of the use and limitations of methods that will prevent or reduce exposure including appropriate engineering controls, work practices, and PPE.
Information on the types, proper use, location, removal, handling, decontamination, and disposal of PPE.
Information on the appropriate actions to take and persons to contact in an emergency involving blood or other potentially infectious materials.
An explanation of the procedure to follow if an exposure incident occurs, including the method of reporting the incident and the medical follow-up should be made available.

An explanation of the signs and color labels required by the 29 CFR 1910.1030.
An opportunity for interactive questions and answers with the person conducting the training session. The person conducting the training shall be knowledgeable in the subject matter covered by the elements contained.

Training records shall include the following:

Names of training attendants.
The dates of the training sessions.
The contents or a summary of the training sessions.
The names and qualifications of persons conducting the training sessions.
Training records shall be maintained for 3 years from the date on which the training occurred.

REVIEW QUESTIONS

True or False

1. T or F—Training records shall be maintained for 5 years from the date on which the training occurred.
2. T or F—Employers must establish and keep a sharp injury log, which became law on April 18, 2001.
3. T or F—When an accident occurs where blood or other potentially infectious materials is found on equipment or services, these items and surfaces should be cleaned and decontaminated.
4. T or F—An employer should make available Hepatitis B vaccine and vaccination series to all employees who have occupational exposure at no cost to the employee.
5. T or F—The safety director of an individual company has the authority to establish a procedure to comply with the Bloodborne Pathogen Standard 29 CFR 1910.1030.

RESOURCES

Academy of Certified Hazardous Materials Managers, PO Box 1216, Rockville, MD 20849, 2007.
J.J Keller & Associates, Inc., PO Box 368, Neenah, WI. 54457
The Grey House safety & security directory. Grey House Publishing, Millerton, NY, 2006.
Occupational Exposure to Bloodborne Pathogens OSHA 3127, 1996.
Occupational Exposure to Bloodborne Pathogens—Precautions For Emergency Responders OSHA 3106, 1998.
Occupational Safety and Health Administration, Bloodborne Pathogens Standard, 1910.1030 U.S Department of Labor, 1998.

Appendix A: Welding

INTRODUCTION

These processes involve the use of either high amp electricity or specialized compressed gases, both of which should be considered high hazard. The potential for fire, explosion, electrical shock, and severe burns are related to these processes. OSHA has established strict guidelines for the protection of workers facing these exposures in the workplace. Standards referenced in this section will be subparts Q and H of CFR 29, OSHA 1910.

Responsibilities/Procedures

Clothing

To protect against flying sparks and hot metal, clothing requires:

At a minimum, shirts should be long sleeved, have no pockets, be worn outside of the trousers, and have the collar buttoned.

Trousers should have no cuffs and extend well down over protective leather shoes.

Personal Protective Equipment

Welding shields or helmets providing full face protection should be worn at all times.

Properly tinted eye wear is required of welder and helper.

Fire-resistant aprons and gloves, usually made of leather, should be worn at all times.

Respirators are required when working in unventilated areas or when working with metals, such as zinc, which is a health hazard.

Ventilation

Welding should never be conducted in unventilated areas, unless employees are protected by air-purifying or fresh-air respirators.

When available, ventilation should be localized or set up to draw fumes away from the breathing zone of the employee.

Fire Prevention

Welding areas should be free of recognizable combustibles.

Adequate fire extinguishers must be readily accessible (within 30 feet).

Fire watches must be posted when hot slag or sparks may fall and contact combustible materials.

Welding Cable and Electrodes

Cables should be securely attached to the welding machine with all exposed metal securely insulated.

Never expose cables to abuse such as vehicle or equipment traffic, since this may damage insulation, exposing energized electrical leads.

Cables should never be run through standing water.

Never change electrodes with bare hands or wet gloves, or when standing on wet floor or grounded surfaces.

Cables with splices within 10 feet of the holder shall not be used.

Damages or worn cables exposing bare conductors should immediately be removed from service or repaired with rubber, plastic, or friction tape, equivalent in insulation to the original cable covering.

Heating and Cutting

Many of the same precautions associated with welding are applicable to heating and burning. The hazard of compressed gas use and storage is discussed in detail in this section, but by no means should it be considered all-inclusive. Refer to OSHA 1910 for additional details on this topic.

Clothing

At a minimum, shirts should be long sleeve, have no pockets, and be worn outside the trousers.

Trousers should have no cuffs and extend over protective shoes.

Personal Protective Equipment

Properly tinted protective goggles should be worn at all times.

For extensive heating and cutting, fire resistant aprons and gloves, usually made of leather, should be used.

Respirators are required when working in unventilated areas or when working with metals coated with zinc, which is a health hazard.

Ventilation

When available, ventilation should be localized or set up to draw fumes away from the breathing zone of the employee.

Fire Prevention

Work areas should be free of recognizable combustibles.

Adequate fire extinguishers must be readily accessible (within 30 feet).

Fire watches must be posted when hot slag or sparks may fall and contact combustible materials.

Compressed Gas Cylinders

All cylinders must be secured in an upright position at all times. Only when "charged out" will exceptions be made.

When in storage, all cylinders must have caps securely in place.

Cylinders in storage should be kept away from sources of heat and should never be stored in unventilated areas.

Cylinders should be stored at least 20 feet from highly combustible materials.

Oxygen and acetylene should never be stored together. A minimum distance of 20 feet or a non-combustible barrier at least 5 feet in height with a fire rating of a least 1 half hour shall separate the cylinders.

Operating Procedures

Valve protection caps shall not be used for lifting cylinders from one vertical position to another.

Unless cylinders are secured in a special truck, regulators will be removed prior to transporting cylinders.

Regulators and gauges must be in good working condition prior to use.

Never use a regulator with damaged gauges.

Open cylinder valves slightly prior to attaching regulators. This "cracking" is to clean dust and debris from the valve.

Supply hoses should not be placed in areas where they may be subject to falling sparks or hot metal.

Before removing regulators, cylinder valves should be closed and gas released from the lines.

Backflow check valves should be installed on all handheld torch sets and at the regulator to minimize possible fires or explosions.

Damaged hoses should be removed from service or repaired immediately.

Never use a lighter or matches to ignite compressed gases. A "striker" should always be used.

Appendix B: Material Handling and General Housekeeping

INTRODUCTION

The definition of materials handling is the lifting, moving, and placing of materials of a variety of shapes. This may be done manually or with different kinds of equipment to make the task easier for the worker. A number of hazards come into play when handling materials and equipment. According to the National Safety Council's publication *Injury Facts* (2008), 20 to 25% of most disabling occupational injuries result from the handling of different materials. Manual materials handling accidents result in a variety of injuries, especially to hands, feet, and legs, with most common injuries.being to backs. The need for special equipment to prevent the injuries associated with lifting heavy loads manually should be a key element of a company's accident and injury records. The General Industry Standard (1910.176) is one of the requirements that define the materials handling section of handling and storing materials. Other specific requirements fall under the non-hazardous section on handling and sorting materials concerned with preventing injuries. A partial listing follows:

- Welding, Cutting, and Brazing (1910.252)
- Pulp, Paper, and Paperboard Mills (1910.201)
- Bakery Equipment (1910.253)

Another form of control involving material transfer is the use of containers to collect overflow spills and the leaking of materials when transferring from one container to another. Safety planning and practices for commonplace tasks need to be as thorough as for operation with unusual hazards.

Commonplace tasks make up the greater part of the daily activities of most employees and, not unexpectedly, offer more potential sources of accidents with injuries and property damage. Every operation or work assignment begins and ends with handling of materials. Whether the material is a sheet of paper (paper cuts are painful) or a cylinder of toxic gas, accident risks can be reduced with thorough planning. Identifying obvious and hidden hazards should be the first step in planning work methods and job practices. Thorough planning should include all the steps associated with good management, from job conception through crew and equipment decommissioning.

Most of the material presented in this section is related to the commonplace and obvious. Nevertheless, a majority of the incidents leading to injury, occupational illness, and property damage stem from failure to observe the principles associated with safe materials handling and general housekeeping.

A less obvious hazard is potential failure of used or excessive motorized handling or lifting equipment.

Responsibilities/Procedures

Lifting and Moving

Lifting and moving of objects should be done by mechanical devices rather than by manual effort whenever this is practical. The equipment used must be appropriate for the lifting or moving task. Lifting and moving devices must be operated only by personnel trained and authorized to operate them. Employees must not be required to lift heavy or bulky objects that overtax their physical condition or capability.

Rigging

Planning for safe rigging and lifting must begin at the design stage, and lifting procedures must be developed for assembly and installation. The lifting procedure should be developed and discussed with employees.

Responsibility for all rigging jobs is shared between the rigging crew and the customer. The customer is responsible for defining and requesting the move, for providing technical information on relevant characteristics of the apparatus including special lifting fixtures when required, for providing suggestions on rigging and moving, and for assigning someone to represent them both in planning and while the job is being carried out.

The riggers are responsible for final rigging and for carrying out whatever moves have been designated. Before any movement takes place, however, each representative must approve the rigging and other procedures associated with the intended move. Each must respect the responsibility and authority of the other to prevent or terminate any action he or she judges to be unsafe or otherwise improper.

The supervisor must make certain that personnel know how to move objects safely by hand or with mechanical devices in the operations normal to the area and must permit only those employees who are formally qualified by training and certification to operate a fork truck, crane, or hoist. The supervisor must enforce the use of safe lifting techniques and maintain lifting equipment in good mechanical condition.

Employees need to observe all established safety regulations relating to safe lifting techniques. A responsible organization provides training programs followed by certification for employees who have demonstrated the ability to operate fork trucks of up to 4-ton capacity and for incidental crane operations that require no special rigging.

Manual Lifting Rules

Manual lifting and handling of material must be done by methods that ensure the safety of both the employee and the material. It should be policy that employees whose work assignments require heavy lifting be properly trained and physically qualified, by medical examination if deemed necessary. The following are rules for manual lifting:

- Inspect the load to be lifted for sharp edges, slivers, and wet or greasy spots.
- Wear gloves when lifting or handling objects with sharp or splintered edges.
- These gloves must be free of oil, grease, or other agents that may cause a poor grip.
- Inspect the route over which the load is to be carried. It should be in plain view and free of obstructions or spillage that could cause tripping or slipping.
- Consider the distance the load is to be carried. Recognize the fact your gripping power may weaken over long distances.
- Size up the load and make a preliminary "heft" to be sure the load is easily within your lifting capacity. If it is not, get help.

If team lifting is required, personnel should be similar in size and physique. One person should act as leader and give the commands to lift, lower, etc. Two persons carrying a long piece of pipe or lumber should carry it on the same shoulder and walk in step. Shoulder pads should be used to prevent cutting shoulders and help reduce fatigue.

To lift an object off the ground, the following are manual lifting steps:

- Make sure of good footing and set your feet about 10 to 15 inches apart. It may help to set one foot forward of the other.
- Assume a bent-knee or squatting position, keeping your back straight and upright. Get a firm grip and lift the object by straightening your knees—not your back.
- Carry the load close to your body (not on extended arms). To turn or change your position, shift your feet—don't twist your back. The steps for setting an object on the ground are the same as above, but in reverse.

Mechanical Lifting

Mechanical devices must be used for lifting and moving objects that are too heavy or bulky for safe manual handling by employees.

Employees who have not been trained must not operate power-driven mechanical devices to lift or move objects of any weight. Heavy objects that require special handling or rigging must be moved only by riggers or under the guidance of employees specifically trained and certified to move heavy objects.

Inspections

Each mechanical lifting or moving device must be inspected periodically. Each lifting device must also be inspected before lifting a load near its rated capacity.

Defective equipment must be repaired before it is used. The rated load capacity of lifting equipment must not be exceeded.

Material-moving equipment must be driven forward going up a ramp and driven backward going down a ramp.

Traffic must not be allowed to pass under a raised load.

The floor-loading limit must be checked before mobile lifting equipment enters an area. Passengers must not be carried on lifting equipment unless it is specifically equipped to carry passengers.

Load Path Safety

Loads moved with any material handling equipment must not pass above any personnel. The load path must be selected and controlled to eliminate the possibility of injury to employees should the material handling equipment fail.

Equipment worked on while supported by material-handling equipment must have a redundant supporting system capable of supporting all loads that could be imposed by failure of the mechanical-handling equipment.

A suspended load must never be left unattended but must be lowered to the working surface and the material-handling equipment secured before leaving the load unattended.

Off Site Shipping

Material being shipped off site must be packed or crated by competent shipping personnel. Boxes, wooden crates, and other packing materials must be safely consigned to waste or salvage as soon as practicable following unpacking.

Truck Loading

All objects loaded on trucks must be secured to the truck to prevent any shifting of the load in transit. The wheels of trucks being loaded or unloaded at a loading dock must be chocked to prevent movement. Dockboards or bridge plates need to comply with 29 CFR 1910.30

General Housekeeping

All areas controlled need be kept in orderly and clean condition and used only for activities or operations for which they have been approved. The following specific rules must also be followed:

- Keep stairs, corridors, and aisles clear. Traffic lanes and loading areas must be kept clear and marked appropriately.
- Store materials in workrooms or designated storage areas only. Do not use hallways, fan lofts, or boiler and equipment rooms as storage areas.
- Do not allow exits, passageways, or access to equipment to become obstructed by either stored materials or materials and equipment that are being used.
- Arrange stored materials safely to prevent tipping, falling, collapsing, rolling, or spreading— that is, any undesired and unsafe motion.
- Do not exceed the rated weight capacity of stored material for both floor and overhead areas. The load limit and the maximum height to which material may be stacked must be posted.
- Place materials such as cartons, boxes, drums, lumber, pipe, and bar stock in racks or in stable piles as appropriate for the type of material.

Store materials that are radioactive, fissile (one that is capable of sustaining a chain reaction of nuclear fission), flammable, explosive, oxidizing, corrosive, or pyrophoric (will ignite spontaneously) only under conditions approved for the specific use by a responsible safety officer. Segregate and store incompatible materials in separate locations (See NFPA 30 for more information on storage of these materials).

Remove items that will not be required for extended periods from work areas and put them in warehouse storage. Call for assistance. Temporary equipment required for special projects or support activities must be installed so that it will not constitute a hazard. A minimum clearance of 36 inches must be maintained around electrical power panels. Wiring and cables must be installed in a safe and orderly manner, preferably in cable trays. Machinery and possible contact points with electrical power must have appropriate guarding. The controls for temporary equipment must be located to prevent inadvertent actuation or awkward manipulation. When heat-producing equipment must be installed, avoid accidental ignition of combustible materials or touching of surfaces above 60°C (140°F). Every work location must be provided with illumination that meets OSHA requirements. Evaluation of illumination quality and requirements is made by a responsible safety officer, but the supervisor of an area is responsible for obtaining and maintaining suitable illumination.

Areas without natural lighting and areas where hazardous operations are conducted must be provided with enough automatically activated emergency lighting to permit exit or entry of personnel if the primary lighting fails.

OSHA Standards for Forklifts

Forklift users must familiarize themselves with and comply with OSHA Standard 29CFR 1910.178 and ANSI B56.1.

Modifications and additions must not be performed by the customer or user without manufacturer's prior authorization or qualified engineering analysis. Where such authorization is granted, capacity, operation, and maintenance instruction plates, tags, or decals must be changed accordingly.

If the forklift truck is equipped with front-end attachments other than factory-installed attachments, the user must ensure that the truck is marked with a card or plate that identifies the current attachments, shows the approximate weight of the truck with current attachments and shows the lifting capacity of the truck with current attachments at maximum lift elevation with load laterally centered.

The user must see that all nameplates and caution and instruction markings are in place and legible. The user must consider that changes in load dimension may affect truck capacities.

Forklift Maintenance

Because forklift trucks may become hazardous if maintenance is neglected or incomplete, procedures for maintenance must comply with ANSI B56. 1 Section 7 and OSHA Standard 29 CFR 1919.178 g.

Forklift Extensions

Maximum efficiency, reliability, and safety require that the use of fork extensions be guided by principles of proper application, design, fabrication, use, inspection, and maintenance. The user must notify a responsible safety officer before purchasing extensions or having them fabricated.

Fork extensions are appropriate only for occasional use. When longer forks are needed on a regular basis, the truck should be equipped with standard forks of a longer length.

Routine on-the-job inspections of the fork extension must be made by the fork lift operator before each use unless, in the judgment of the supervisor, less frequent inspections are reasonable because of his or her knowledge of its use since the last inspection. Extensions must be inspected for evidence of bending, overload, excess corrosion, cracks, and any other deterioration likely to affect their safe use. All fork extensions must be proof load tested to establish or verify their rated capacities. A load equal to the rated capacity of the pair at a particular load center multiplied by

1.15 must be placed on each fork extension pair and fork assembly and supported for a period of 5 minutes without any significant deformation.

Rated capacity must be determined at significant load centers, including the midpoint of the extension and at the tip. Once determined, the rated capacity and load center information must be shown by stamping or tagging the extensions in a protected location of low stress. The proof load test must be witnessed by a mechanical engineer or designer. Whenever evidence of deterioration is detected or whenever the extensions have been overloaded, magnetic particle inspection must be performed.

Safety Inspection

Each operator is responsible for the safety and safety inspection of his or her lifting devices (such as screw pin shackles, hoist rings, commercial equipment, etc.) and for its lifting fixtures (such as spreader bars, special slings, equipment, etc.). All lifting fixtures designed by your own company must be proof tested to twice their maximum rated loads before they are placed in service. A magnetic particle inspection or other appropriate crack detection inspection is required after the proof test. The capacity must be marked on the lifting fixture so that it is clearly visible to the equipment operator. All lifting device pins 2 inches in diameter or larger must have a magnetic particle inspection before they are placed in service.

All lifting fixtures must be inspected at least once every 4 years (or upon request), using magnetic particle detection or other appropriate methods.

BIBLIOGRAPHY

Godish, Thad. *Air quality*, 4th ed., CRC Press, Boca Raton, FL, 2004.

Mulhausen, John R. and Damiano, Joseph. *A strategy for assessing and managing occupational exposures,* 2nd ed., American Industrial Hygiene Association. Fairfax, VA. 1998.

29 CFR, 1910.119, *Process safety management of highly hazardous chemicals.*

29 CFR 1910.1200.

Appendix C: Transportation Safety

INTRODUCTION

At the direction of the U.S. Congress under the Commercial Motor Vehicles Safety Act of 1986, national standards were developed for every individual driving commercial motor vehicles in the country. The regulations (49 CFR Part 383) were finalized by the Federal Highway Administration on July 1, 1988).

These regulations establish uniform guidelines for licensing and testing of drivers that all states must use as a basis of issuing licenses to drivers domiciled in their states. All drivers of commercial motor-carrier vehicles must pass written and skill tests demonstrating their knowledge and ability to drive a commercial motor vehicle. After that date no one in the United States may operate a commercial motor vehicle unless they possess a valid commercial drivers license (CDL).

Definitions

CDL: Commercial Driver's License

Off-Road Vehicle: Any motor vehicle that is utilized away from an interstate road to transport machinery, compost, etc., for company tasks, but whose operator does not require a CDL.

DOT: Department of Transportation
FMCSR: Federal Motor Carrier Safety Regulations

Responsibilities/Procedures

Management

It is the responsibility of management to ensure the following rules and regulations and recommended practices are complied with to maintain a comprehensive transportation safety compliance program.

Policy statement issued by management:
  Distributed to all drivers
  Prominently posted
  Recently updated

Vehicle Fleet Safety Manager

It is the responsibility of the vehicle fleet safety manager to ensure that the following rules are followed:

Clearly defined authority and responsibility
Adequate time and resources to perform duties
Cognizant with safety rules and Department of Transportation (DOT) regulations
Employees hold a CDL

Administrative

Approve route and other restriction (height and weight clearances, etc.).
Record hours of service (where/when applicable).
Track insurance claims and lawsuits.
Perform accident analysis and causation.
Inspect vehicles on a scheduled basis.

Safety

Distribute general safety rules to all drivers.
Record and maintain records of signed receipts stating that all employees have acknowledged stated safety rules.
Update and review rules periodically.
Develop discipline system for chargeable accidents or violations.

Driver Selection Process

Perform motor vehicle record (MVR) check and ensure current CDLs are held by employees.
Verify all driving records with prior employers.
Maintain written test (CDL).
Develop driving test in the vehicle type new employee will be driving.

Interview new employees, asking questions relating to their general understanding of safety rules.

Alcohol and Drug Testing Program

Develop a written drug and alcohol policy statement.
Maintain a documented drug and alcohol testing program.
Test drivers prior to employment.
Ensure that the drug and alcohol program meets the Department of Transportation (DOT) specifications.

Criteria for driver incentive awards should include:

Attendance
Timeliness of on-the-road tasks
Preventable accidents
Moving violations
Accuracy of log book entries (where applicable)
Accuracy of pre- and post-trip vehicle inspections
Adherence to general safe driving rules

Post Accident Procedures

Accident investigation should be conducted.
Containment and clean up of fuel and or cargo (where applicable).
Reporting procedures to fleet safety manager.

Vehicle Management

Selection and modification of vehicles and equipment
Scheduled vehicle replacement
Preventive maintenance program
Driver compliance reports and follow up
Periodic inspection and maintenance of non-regulated and off-road vehicles (where applicable)

Employees

It is the responsibility of all employees to comply with all rules and regulations set by management of the company in which they are employed. In the event of any disregard for company rules and regulations, disciplinary actions should be enforced; depending on the violation, possible termination of employment could be a possibility.

Record Keeping

Driver Qualifications File

There should be a file maintained for every driver employed containing the following information and documents:

- Driver qualification check list
- Application for employment
- Pre-employment urinalysis notification
- Employment eligibility verification
- Certification of compliance
- Request for information from previous employer
- Road test record and certificate
- Written examination and certificate
- Physical examination and certificate
- Driver's date sheet
- Record of violations
- Annual review of driver record
- Copy of pocket cards
- Statement of responsibility
- Copy of current CDL
- Copy of substance abuse awareness certificate

Training/Certification

Some good times to perform driver training classes are:

For new drivers
After accidents and moving violations
On new or modified equipment
Periodic refresher courses for all drivers

Justification

Commercial Motor Vehicles Safety Act of 1986
49CFR Part 391
49CFR Part 392
49CFR Part 393
49CFR Part 395
49CFR Part 396 and
49CFR Part 397 of the FMCSR
DOT Regulations

Appendix D: Overhead Hoists and Slings

INTRODUCTION

Material handling can be very dangerous if the proper equipment is not used or if it is in poor condition. When dealing with hoists and slings, the danger is increased due to loads being elevated, possibly overhead. When a failure occurs with elevated loads, employees can be very seriously injured or even killed. Every effort should be made to ensure that equipment is in good condition prior to each use.

RESPONSIBILITIES/PROCEDURES

Overhead Hoists

The safe working load, as determined by the manufacturer, will be indicated on the hoist and must never be exceeded.

The supporting structure to which the hoist is attached must have a working load equal to that of the hoist.

The hoist support shall be arranged to allow for free movement of the hoist and should not restrict the hoist from lining itself up with the load.

Hoist operators should stand clear of suspended loads at all times.

Hoists should be installed, tested, inspected, and maintained in accordance with manufacturer's recommendations.

Mechanized overhead hoists must be inspected and certified annually to ensure proper working order.

Slings

Slings may be steel chains, wire rope, or synthetic web.

Slings with visible damage shall not be used.

Slings should never be loaded in excess of their capacity, which should clearly be marked on each sling.

Slings should be padded or protected from the sharp edges of their load.

Employees shall be kept clear of loads about to be lifted, and of suspended loads.

Never place hands or fingers between the sling and its load while the sling is being tightened around the load.

Each day prior to use, the sling and all attachments shall be inspected for damage or defects.

Wire rope slings should immediately be removed from service if any of the following conditions exist:

1. There are ten randomly distributed broken wires in one rope lay, or five broken wires in one strand in one rope lay (one section of wire cable).
2. There is wear or scraping of one-third of original diameter of outside individual wires.
3. There is kinking, crushing, bird caging, or any other damage that restricts hoist because it is not in line with the load.
4. There is evidence of heat damage.
5. End attachments are cracked, deformed, or worn.
6. Hooks have been opened more than 15% of the normal throat opening.
7. There is corrosion of the rope or end attachments.

Synthetic web slings should immediately be removed from service if any of the following conditions exist:

1. Acid or caustic bums
2. Melting or charring of any part of the sling surface.
3. Visible snags, punctures, tears, or cuts
4. Broken or worn stitching
5. Visible "red threads," usually associated with cutting or tearing, indicating damage

JUSTIFICATION

29 CFR 1910.179

29 CFR 1910.184

Appendix E: Machine Guarding and Portable Tools

INTRODUCTION

Based on the fact that since the industrial revolution machines, designs and materials have changed, even though the principles of guarding has been in effect for a very long time. Accident data from the National Safety Council's publications show with the growth of more machine injuries resulting from machines have increased over the years. Specifically designed guards on machines are intended to keep workers and their clothing from coming into contact with these machines. Well placed guards must not create any further hazards and keep the workers from reaching into dangerous parts of a machine.

Definitions

Cutting Actions: Any machine that cuts or removes material has cutting actions. The area of a machine's cutting action is the point of operation. Some cutting actions include lathes, grinders, saws, milling machines, and shaping machines.

Hand Tools: Handheld devices that manually assist the employee in a work operation. Examples of hand tools are: wrenches, pliers, snips and cutters, knives, axes, hammers, screwdrivers and portable electric, pneumatic, and gas-powered tools.

Pneumatic Tools: Are powered by compressed air. The air pressure is often high enough to cause serious injury if proper safety precautions are not taken.

Ground Fault Circuit Interrupter: It is designed to trip much faster than an ordinary breaker should a fault to ground occur, cutting off power to the tool quickly enough to prevent electrocution. When a worksite is wet, or when the operator is standing on metal, a GFCI is placed in line with the tool.

Guarding: Any means of effectively preventing personnel from coming in contact with the moving parts of power tools and machinery that could cause physical harm.

Point of Operation: The area on a machine where material is positioned for processing or change; that point at which cutting, shaping, boring, or forging is accomplished on stock.

Power Transmission: All mechanical components, including gears, camshafts, pulleys, belts, and rods that transmit energy and motion from the source of power to the point of operation.

Ingoing Nip Point: A hazard area created by two or more mechanical components rotating in opposite directions in the same place and in close conjunction or interaction.

Responsibilities/Procedures

The safety manager is responsible for issuing the policy for portable tools and machine guarding to each operating personnel and updating the policy as necessary to ensure compliance with the most current regulations.

The company's safety department shall administer this program and ensure audit procedures are in place to provide a workable program.

The safety manager shall be responsible for communicating this policy to the work force and enforcing its use.

Each employee, salaried or hourly, should be responsible for complying with this policy.

Machine Controls And Equipment

A mechanical or electrical power control shall be provided on each machine to make it possible for the operator to cut off power to the machine without leaving his or her position at the point of operation.

On machines driven by belts and shafting, a locking-type belt shifter or an equivalent positive device shall be used.

On machines where injury to the operator might result if motors were to restart after power failures, provisions shall be made to prevent machines from automatically restarting upon restoration of power.

Power controls and operation controls shall be located within easy reach of the operator while at his or her regular work location, making it unnecessary to reach over the cutter to make adjustments. This does not apply to constant pressure controls used only for setup purposes.

On machines operated by electric motors, positive means shall be such that controls or devices are rendered inoperative while repairs or adjustments are being made to the machines they control (i.e., unplugging machine, shutting off circuit breaker).

Each operating treadle shall be protected against unexpected or accidental tripping.

There shall be ample work space around the machine, as required by the type of operation.

The operator shall make an inspection of the machine prior to each start. This should include a check of operational controls and safety devices, including all guards. Any defects must be reported to the operator's immediate supervisor. The defective machine is to be taken out of service and repaired before it is used again.

Moving Parts of Equipment

Belts, gears, shafts, pulleys, sprockets, spindles, drums, fly wheels, chains, or other reciprocating, rotating, or moving parts, nip points, shear points, crush points, and trapping spaces of equipment shall be guarded if such parts are exposed to contact by employees or otherwise pose a potential hazard.

Disconnect Switches

All fixed power-driven woodworking and metal working tools and machines shall be provided with a readily accessible disconnect switch that can be either locked or tagged in the off position.

Old vs. New Equipment

Old as well as new machines and equipment must meet the requirements of this section. If equipment has been purchased without the required guarding or the guarding has been removed, the equipment must be tagged out of service until the appropriate guarding is provided.

Usually, the most practical approach for the purchase of new equipment is to specify all the appropriate guarding to be provided by the manufacturer, and to ensure the guarding is maintained on the equipment for use at all times.

Guarding Horizontal Shafting

All exposed parts of horizontal shafting 7 feet or less from floor or work levels shall be protected by a stationary casing enclosing shafting casing, or by a trough enclosing sides and top or sides and bottom of shafting, as the location requires.

Guarding Vertical and Inclined Shafting

Vertical and inclined shafting 7 feet or less from floor or working levels shall be enclosed with a stationary casing.

Protecting End Shafts

Shall present a smooth edge and end, and shall not project more than one-half inch the diameter of the shaft unless guarded by non-rotating caps or safety sleeves.

Guarding Pulleys

Pulleys, any part of which are 7 feet or less from the floor, shall be guarded.

Broken Pulleys

Pulleys with cracks or pieces broken out of rims shall not be used.

Belts, Rope and Chain Drives

Where both runs of horizontal belts are 7 feet or less from the floor level, the guard shall extend to at least 15 inches above the belt or to a standard height except that where both runs of a horizontal belt are 42 inches or less from the floor, the belt shall be fully enclosed.

Vertical and inclined belts shall be enclosed by a guard arranged in such a manner that a minimum clearance of 7 feet is maintained between the belt and the floor at any point outside the guard.

Gears

Gears shall be guarded by complete enclosure. A standard guard at least 7 feet high, extending 6 inches above the mesh point of the gears, or a band guard covering the face of the gear and having flanges extended inward beyond the root of the sides. Where any portion of the train or gears guarded by a band guard is less than 6 feet from the floor, a disk guard or complete enclosure to a height of 6 feet shall be required.

Sprockets and Chains

Sprockets and chains shall be enclosed unless they are more than 7 feet above the floor or work station. Where the drive extends over other work areas, protection against falling objects shall be provided.

Couplings

Couplings shall be so constructed as to present no hazard from bolts, nuts, set screws, or revolving surfaces. They shall be covered with a safety sleeve or guarded by an enclosed guard.

Grinding Tools

Abrasive wheels shall fit freely on spindles. They shall not be forced, nor shall they be too loose. Before the grinding wheels are mounted, they shall be inspected and given the "ring" test to ensure they have not been damaged.

Before mounting the grinding wheel, the speed of the machine shall be checked to ensure that it does not exceed the maximum operating speed marked on the grinding wheel. Safety guards shall be adjusted to allow maximum protection for the grinding wheel operator. Grinding wheels shall be stored in a clean and dry location.

Portable Belt Sanding Machines

Belt sanding machines shall be provided with guards at each nip point where the sanding belt runs onto a pulley. These guards shall prevent the hands and fingers of the operator from coming in contact with the nip points. The unused run of the sanding belt shall also be guarded to prevent accidental contact.

Portable Circular Saws

All portable, power-driven circular saws having a blade diameter greater than 2 inches shall be equipped with guards above and below the saw base plate or shoe. The upper guard shall cover the saw to the depth of the teeth, except for the minimum arc required to permit the base to be tilted for bevel cuts. The lower guard shall cover the saw to the depth of the teeth, except for the minimum arc required to allow proper retraction and contact with the work. When the tool is withdrawn from the work, the lower guard shall automatically and instantly return to the covering position.

Portable Band Saws

Guards shall be provided at each pulley and extended along the non-cutting portion of the band saw. Because a guard is not feasible along the cutting portion, proper hand position while using the tool is imperative. Both hands are necessary to handle this type of saw, and shall be placed into the provided handles.

Pneumatic Tools

A tool retainer shall be installed on each piece of equipment that, without such a retainer, may eject the tool. The hose and hose connections used for conducting compressed air to pneumatic equipment shall be designed for the pressure and service to which they are subjected.

The hose shall be secured to the power tool by some positive means that will prevent the hose from becoming accidentally disconnected. Hose connections shall be secured with a quick disconnect, or with a pin or wire.

All pneumatically driven nailers, staplers, and other similar equipment provided with an automatic fastener feed and operating at more than 100 pounds per square inch (psi) pressure at the tool shall have a safety device on the muzzle that will prevent the tool from ejecting fasteners when the muzzle is not in contact with the work surface.

Spray Guns

Airless spray guns of the type that atomize paints and fluids at high pressures (1,000 psi or more) shall be equipped with an automatic or visible manual safety device. This will prevent pulling of the trigger and releasing of the paint or fluid until the safety device is manually released. In lieu of the above, a diffuser nut to prevent high-pressure, high-velocity release while the nozzle tip is removed, plus a nozzle tip guard to prevent the tip from coming into contact with the operator, or equivalent protection, shall be provided.

Woodworking Machinery

All woodworking machines (i.e., band saws, table saws, drill presses, and lathes, etc.) shall be constructed and maintained so that, while running at full or idle speed and with the largest cutting tool attached, they are free of excessive noise and harmful vibrations.

All belts, blades, shafts, gears, and other moving parts shall be fully enclosed or safeguarded so that the worker cannot accidentally come in contact with them. If there are moving parts in back of the machine (i.e., the side away from the worker), these parts shall be covered or the area closed to prevent entry.

Because most woodworking operations involve cutting, it is often difficult, although necessary, to provided guards at the point of operation. On most machines, the point of operation guard should be movable to accommodate the size and shape of the wood, balanced so as not to impede the operation, and strong enough to provide protection to the operator.

All machines, except portable or mobile ones, shall be securely fastened to the floor or other suitable foundation to eliminate all machine movement or "walking." There must be ample work space around the machine, as required by the type of operation.

The operator shall make an inspection of the machine prior to each start. This will include a check of operational controls, safety guards and devices, power drives, sharpness of cutting edges, and other parts that are to be used. Any defects shall be reported to the operator's immediate supervisor. Defective machines shall be taken out of service and repaired before being used again (see lockout/tagout procedures).

For a tool guarding procedure to be effective, the employees must understand why guards are necessary. Precautionary explanations along with periodic reminders shall be given so that employees respect the value of machine guarding. Follow-up measures, in the form of proper supervision, regular inspections, and accident investigations are also necessary to assure compliance with machine guarding regulations and practices.

Total safety in the use of tools includes proper worker positioning and use of adequate PPE for each task.

BIBLIOGRAPHY

ANSI B 173.3-1985. *Safety requirements for heavy striking tools.* American National Standards Institute, New York.

ANSI B 173.1-1992. *Safety requirements for nail hammers.* American National Standards Institute, New York.

ANSI B 7.1-1982. *Safety requirements for the use and care of abrasive wheels.* American National Standards Institute, New York.

ANSI Z87.1-1989. *Protection for occupational and educational eye and face protection.* American National Standards Institute, New York.

Blundell, J.K. *Safety engineering: Machine guarding accidents.* Hanrow Press. Del Mar, CA, 1987.

Hand tool safety: Guide to selection and proper use. Hand Tool Institute, White Plains, NY, 1976.

Power Press Safety Manual, 5th ed. National Safety Council, Itasca, IL, 2002.

Index

D

E